Investigating Ancient Tillage

An experimental and soil micromorphological study

Helen Lewis

BAR International Series 2388
2012

Published in 2016 by
BAR Publishing, Oxford

BAR International Series 2388

Investigating Ancient Tillage

ISBN 978 1 4073 0980 4

BAR Publishing is the trading name of British Archaeological Reports (Oxford) Ltd. British Archaeological Reports was first incorporated in 1974 to publish the BAR Series, International and British. In 1992 Hadrian Books Ltd became part of the BAR group. This volume was originally published by Archaeopress in conjunction with British Archaeological Reports (Oxford) Ltd / Hadrian Books Ltd, the Series principal publisher, in 2012. This present volume is published by BAR Publishing, 2016.

Printed in England

BAR
PUBLISHING

BAR titles are available from:

BAR Publishing
122 Banbury Rd, Oxford, OX2 7BP, UK
EMAIL info@barpublishing.com
PHONE +44 (0)1865 310431
FAX +44 (0)1865 316916
www.barpublishing.com

Table of Contents

Acknowledgements

This volume is based on PhD research carried out at the Charles McBurney Geoarchaeology Laboratory, Department of Archaeology, University of Cambridge. I would like to Vladimir Olkhovoi, Natalya Rostovsteva, Vladislav Stukalov, the British Federation of Women Graduates, Lejre Historisk-Arkæologisk Forsøgscenter, the Garrod Fund, the Board of Graduate Studies and Queens' College, Cambridge, for financial assistance. I am very grateful to my referee for his advice and suggestions.

I would also like to thank my family for their support, as well as Marie-Louise Stig Sørenson, Donald Davidson, Charles French, Richard Macphail, Martin Jones, Anne Gebhardt, Marianne Rasmussen, Bjarne Grønnow and the staff at Lejre, Gordon Spoor, Jacques Desbiolles, Roy Newland and Pete Howsam at Silsoe College, Cranfield University, Butser Ancient Farm, Julie Boreham, Clare Ellis, Jen Heathcote, Chris Stevens, Gillian Wallace, Manuel Arroyo-Kalin, Thomas Beckmann, Melissa Goodman-Elgar, Karen Milek, Gwil Owen, Wendy Matthews, and the Pitt-Rivers Laboratory. I am also grateful to the Thy Project and Als Project, Jens-Henrik Bech, Kristian Kristiansen and Tim Earle.

This volume was (finally) written up at University College Dublin School of Archaeology; I am grateful to Conor McDermott, Rob Sands and Tadhg O'Keeffe for advice and assistance. I would like to thank the various people who periodically emailed (read: nagged) me for a copy of my thesis – while the detailed data and many images are not presented here and some aspects are slightly out of date, I hope that the complete discussion of the experimental results will go some way to meeting those requests.

Finally, I am particularly grateful to the late Peter Reynolds and Andrew Sherratt for their encouragement, suggestions and support, and for enthusiastically sharing my interest in early farming (if not my approach to it), and I would like to dedicate this volume to both of their memories.

Chapter 1 Introduction

In an article reviewing the application of soil micromorphology to studies of ancient arable farming it was suggested that '...soil micromorphology itself cannot be used at present to provide definitive evidence for cultivation in ancient soils' (Carter & Davidson 1998, 545). This volume addresses this statement by presenting the results of experimental research on the identification and characterisation of ancient cultivation using a soil micromorphological approach.

Soil micromorphology is a method of analysis that involves examining soil structure and components in their undisturbed state (Courty *et al.* 1989, xvii). It was developed in the early 1900s (Kubiena 1938, 70-4), and has been widely utilised in pedology and agronomy (e.g. Kooistra *et al.* 1996; Kooistra 1987; Bullock 1983); the history of the approach in regard to tillage in agronomy has been discussed by Jongerius (1983) and Kooistra *et al.* (1996). Much work has focused on soil fertility and organisational features relating to drainage systems and amendment (e.g. Borchert 1964; 1967; Altemüller & Banse 1964; Turelle & McCalla 1961; Pagliai *et al.* 1983), and on the impact of tillage implements on soils, particularly in regard to how low-impact tillage could reduce erosion and loss of fertility (Jongerius 1970; Bouma 1969). This body of work has addressed means of sustaining soil structural integrity and fertility (e.g. Kooistra 1987; Darmody & Norton 1994; Pagliai 1987; 1994), and has assessed land-use impact on soils (e.g. Jongerius & Jager 1964; Pawluk 1980; Kooistra *et al.* 1985; 1990; Barratt 1968; 1971).

The history and applications of soil micromorphology in archaeology are discussed by Goldberg (1980; 1983), Goldberg and Macphail (2006, 354-67), French (2003), Courty (1992), Courty *et al.* (1987; 1989) and Macphail *et al.* (1990b), among others. Soil micromorphology is an especially appropriate means of examining past cultural events which interacted directly with the soil, such as digging and tillage (Gebhardt 1992, 373). Reviews of the application of soil micromorphology to the study of ancient cultivation have been presented by Carter and Davidson (1998) and Macphail (1998a).

The idea that one might be able to recognise signatures of ancient arable activity in modern soil profiles has been mooted in archaeology since at least the 1950s, when Dalrymple (1958) and Cornwall (1958; 1963) suggested that field-based study of buried soils could allow interpretation of past environment and land use, including the identification of ancient grasslands and cultivated soils. It is important to identify and characterise ancient agricultural practices from localised remains, so as to explore land-use histories in relation to specific archaeological sites, phases, and landscape features such as bounded field systems, in a way that is impossible from landscape-scale information such as pollen. Soil micromorphological research has explored fundamental land-use issues, aiming to distinguish between ancient arable and pasture fields, and between these and 'natural' grasslands or forested areas (e.g. Gebhardt 1990; Macphail *et al.* 1987; 1990a). Attempts to identify and discuss ancient land use have focused greatly on the impact of tillage, especially distinguishing between various cultivation implements (Gebhardt 1990; Lewis 2002), and on identifying land-use systems and changes in these over time, such as comparing slash-and-burn and fallow systems (e.g. Romans & Robertson 1975; Gebhardt 1993; 1995; Macphail 1986a; 1998a; Macphail *et al.* 2000; Goldberg and Macphail 2006). There has also been a great deal of work on characterising specific arable practices, especially manuring (e.g. Davidson & Carter 1998; Dockrill & Simpson 1994; Macphail *et al.* 1987). In these cases, the basis for recognising ancient arable farming from undisturbed soil samples in thin section is the identification of the physical influence of cultivation, through a combination of structural and fabric characteristics, including mixing of horizon materials, the presence of angular or fragmented aggregates created by implement action, or evidence for an interruption or even absence of expected soil properties. In some cases the results of complimentary geochemical studies have been applied to support interpretations of ancient cultivation and/or amendment (e.g. Goldberg & Macphail 2006; Macphail & Crowther 2008; 2009; Macphail *et al.* 2000; 2004).

The soil disturbance created by tilling implements has specific, immediate, and relatively easily recognisable macroscopic effects, so tillage should in theory be among the simplest of arable practices to study microscopically. Tillage is an activity that is directly related to arable farming: if prehistoric tillage indicators are found, a contemporaneous knowledge of arable practices may be inferred. Soil micromorphology has proven to be most useful in detecting ancient tillage indicators in buried soil horizons (Gebhardt 1992, 373), such as those under monuments, alluvium, colluvium, and aeolian deposits. Much archaeological work regarding the identification and understanding of tillage has consisted of interpretation of known tilled horizons (e.g. Romans & Robertson 1983b; Macphail 1986a; Lewis 2002) or horizons associated with other environmental archaeological evidence of cultivation (e.g. pollen, molluscs, seeds).

Gebhardt's (1990; 1992; 1995; 1996) detailed soil micromorphological study of cultivated land compared microscopic soil characteristics identified from agronomic experiments using modern tools (after Jongerius 1970; 1983; Kooistra 1987) to those produced by reconstructed ancient tools, including ards, spades and hoes. Experimental tillage with replicas of these prehistoric tool types produced recognisably distinctive structures in soil horizons (Gebhardt 1990; 1991b; Macphail *et al.* 1990a, 61). Comparison to archaeological samples showed that some of the experimentally-produced microscopic features could be identified in ancient cultivated buried soils. Most of the observed indicators were, however, restricted in their interpretative capacity regarding specific implements, since they were mainly markers of general soil disturbance and of burning, which could be caused by activities other than cultivation. In Gebhardt's pilot study and later field applications (e.g. Gebhardt 1995; Davidson & Carter 1998), the micromorphological line of enquiry proved to be most productive in identifying ancient arable land use in buried soils when associated with other indicators of cultivation (e.g. ard marks, pollen of cultigens, historic evidence).

It has been suggested that the true promise of soil micromorphology lies in the potential to identify ancient land use in the absence of such markers (Bryant & Davidson 1996; Macphail *et al.* 1990a, 53), and that this might be possible upon establishment of a more extensive and detailed micromorphological database. Much commentary has, however, been less optimistic, especially regarding ancient implement indicators (Courty 1996; Courty *et al.* 1996; Carter & Davidson 1998; Davidson & Carter 1998; Usai 2001). Extant soils are demonstrably different to those present in the same locations in prehistory; soils are and always have been extremely complex living systems, linked to specific local characteristics and changing continuously, but at varying rates, over time. Most of the cultural activity in an arable or pastoral soil system occurs in the living topsoil, and most traces of this have seen substantial reworking since ancient times. As such, frequent recourse to modern pedogenetic theory and a certain amount of imagination are required when reconstructing ancient farming activities from soil thin sections. This is particularly true when one's approach is based primarily on soil horizon characteristics, as is the case in most of the archaeological research cited above.

However, it is also possible to investigate microscopic tillage indicators through a more standard archaeological approach, by comparing tillage macro- and microfeatures on an archaeologically typological basis. The origin and development of morphologies specific to archaeological features like ard marks might then be discussed profitably with regard to archaeological and soil formation processes; such a typological approach, describing the characteristics of tillage features themselves, may prove to be less ambiguous, and of more use in relating microscopic indicators to macroscopic archaeological features and landscapes. The micro-structure and organisation of tillage features have rarely been systematically studied archaeologically, and soil micromorphology has primarily been utilised as a means of identifying land use in the absence of such features.

Individual tillage mark features have specific organisational characteristics which can be defined micromorphologically, and relics of these microscopic traits (e.g. panning, compaction, loosening, turning indicators) are as likely to survive archaeologically as are tilled horizon indicators (structure, organic and other inclusions, manuring indicators, etc). Indeed, the various density zones created in and around implement marks may form the basis for many tilled horizon characteristics, and may be important in soil profile changes involved in horizon boundary creation. For instance, the structural changes related to internal slaking (Jongerius 1983, 122; Pawluk 1980) appear to have a significant and long-lasting influence on soil profile development. The study of distinctive aspects of tillage mark features is the subject of this volume.

The significance of land use

Land use refers to the utilisation of an area of the earth; it also signifies cultural activity that interacts with the physical earth. There are many types of land use, encompassing all aspects of settlement, agriculture, resource exploitation, recreational and spiritual activities, where these occur in specific physical places. Land uses may change frequently over time or may be sustained for millennia, depending on many issues, including desire, need, access, availability, and environmental change. Under this broad definition, archaeological sites of all types can, if so desired, be classified through interpretation (e.g. cemetery, dwelling, meeting place, clearing, field), and characterised typologically. Archaeologists do not often discuss things in terms of land use *per se* (which is mainly the remit of planning authorities), but we are frequently engaged in recording and interpreting land-use decisions over long time periods, particularly in settlement and landscape archaeology.

This volume focuses on agricultural land-use practices, specifically arable land use, and in particular the remains of ancient cultivation activities. Arable agriculture has been seen as one of the most important food-procurement land-use types in archaeology, traditionally linked to the concept of the 'agricultural revolution', often seen as key in the development of modern societies (Goody 1977, 7-8; Clark 1952; Schama 1995; Duncan 1993; Frazer 1925, 129), and is often described in terms of crop domestication, economic change, and the 'taming' of the land.

Arable land use is linked in modern perceptions to the conquest of Nature by Culture; indeed, cultivation is one of the key symbols of this taming of the wild, related to a sedentary lifestyle, forming core concepts in proposed Neolithic revolutions in land use (e.g. Hodder 1990; Thomas 1991; 1999; Cooney 1997; Jones 1997; Zvelebil 1986; Zvelebil & Dolukhanov 1991). The dichotomy between wild and tame is an important construct in certain landscape and settlement archaeology studies, particularly those dealing with the beginnings of the Neolithic. The Neolithic as a whole is undergoing considerable revision in archaeology at the moment, at least as a set of global concepts in human history. For instance, the relationship between early villages and early farming is disputed in some areas, the dating of Neolithic domesticates is being moved back into the Mesolithic or Palaeolithic in many places, and the relationship of staple artefact types, like pottery, with localised onsets of agriculture appear to be extremely variable. Nevertheless, the ancient development of agricultural technologies and the spread to and/or influence of these upon non-agricultural societies remain major issues in prehistory. The adoption and development of agriculture was obviously 'revolutionary' in many places and for many ancient societies, whether or not it came as a package with all the Neolithic trappings, or piecemeal and over relatively long time periods. Later developments in agricultural practice, and the interpretation of the world through agriculturalists' eyes, are also linked with profound changes and long-lived traditions in many ancient societies around the world.

In regard to ancient arable land use in particular, the work of Boserup (1965; 1970) and Goody (1976) has been especially influential. For the purposes of this volume, it is sufficient to say that ethnographically-based models have related agricultural land-use practices and technology, including tool types, to organisation of land tenure, the sexual division of certain aspects of labour, levels of social stratification and inheritance patterns in anthropologically documented groups (*ibid.*). These correlated aspects of society have been suggested to be evolutionary (hunting/gathering to farming to industrialisation) (*ibid.*), and have been used archaeologically to interpret ancient social organisation and changes in this over time, despite the fact that the archaeological record in many places demonstrates that these relationships are neither simple, nor easily applicable to all prehistoric societies. There are good indications, however, that various agricultural land uses did have significant meaning in ancient societies, and finding ways to interpret ancient perceptions of land use are thus vital. One example examining the problems of applying ethnographic models to ancient landscapes is that of ancient Wessex, England (French *et al.* 2007; Lewis 1998; 2010), where despite detailed archaeological land-use investigations it has thus far proven impossible to demonstrate a middle Bronze Age transition in agricultural land use that has been interpreted from settlement and monumental remains using these ethnographic models (e.g. Barrett 1994). Since arable land use is a fundamental part of such models, and as it affects the physical earth in substantial ways, assessing theories of social change through archaeological field and laboratory investigations of arable land-use practices is highly desirable (see, e.g., Lewis 2008; 2010).

Arable farming is a land-use activity that takes place mainly in fields, defined here as temporary or permanent clearings for agriculture, and entails the use of cultivation implements to produce arable crops, such as wheat, fodder grasses and pulses. Under the Boserup (1965; 1970) model mentioned above, which is based mainly on length of fallow time, implements (as part of tillage technology) are directly related to specific systems. For instance, and extremely simplistically, shifting forest fallow and bush fallow systems – both of which have been related to burning, low land tenure, polygamy/polygyny, bridewealth and female tillage labour – generally differ in technology as well as fallow length: Boserup's study suggested that forest fallow is based on axes and digging sticks, while bush fallow also involves hoes for weeding. Grass-fallow systems, which have been related to high land tenure, dowry and male tillage labour, along with the presence of herbivores, cereal and fodder crops, and relations with nomadic groups, are said to require some form of plough. Short grass-fallow systems are similar, but are also related to manuring practices (*ibid.*). Goody (1976) suggested that low tenure systems also tend to show low social stratification and to be matrilineal, while high tenure systems are politically complex and tend to show high population densities.

These models are much more complex than presented here, where the purpose is simply to indicate the centrality of perceptions regarding arable land use and tillage implements to many broad anthropological and archaeological interpretations of long-term social change, and they have been criticised from many perspectives by geographers, ethnographers and anthropologists (e.g. Grigg 1979; Hunt 2000; Datoo 1978; Netting 1993; Bender 1978; Bray 1986; Sutter 1987; Gibson 1988). While these models suggest that there may be 'universal' relationships between land use, settlement, technology, social organisation and landscapes, we must be circumspect about simple application of them in archaeology (see Bradley 1972; McGlade 1995; Morrison 1994; Lewis 1998; 2008; 2010; French *et al.* 2007); instead of relying on coincidences in ethnographic data and our perceptions of evolutionary landscape changes, it seems preferable to explore the links between land use and landscape using the archaeological record itself. One part of the evidence required to do this is that which produces information on land-use patterns over time, and it is in this regard that approaches such as soil micromorphology can engage with broader archaeological interpretations and theory.

The scope of this volume

This volume presents a series of experimental investigations designed to explore the identification and characterisation of ancient arable farming through a feature-based morphology approach, and to assess previous work regarding the ability of soil micromorphological approaches to identify ancient tilled soils on the basis of profile and horizon characteristics. Tillage affects many aspects of soil in the long-term, and may be studied using several methods archaeologically (e.g. soil texture, relative phosphate content, organic content and other chemical characteristics); indeed, it is a set of activities best studied through a combination of many approaches which address different types and scales of land-use impact and significance. This volume is based on a targeted PhD study with the aim of investigating the soil micromorphological characteristics of the features of ancient tillage, and that limitation means that less attention is given here to other soil characteristics which are also used to interpret ancient land-use practices.

After an introduction into the types of features created by ancient tilling implements, soil micromorphological investigations are presented from northwest European experiments in reconstructed ancient tillage. While the focus of the volume is on experimental studies, a comparative discussion of a number of archaeological examples links the experimental archaeology to real archaeological situations, and a scheme for characterising ancient implement marks is outlined.

Both laboratory and field experiments were conducted for this project (Table 1). Laboratory experiments were carried out using a reconstructed Donneruplund ard in the soil bin at Silsoe College, Cranfield University, England, and field experiments using another Donneruplund replica and a Døstrup ard replica were sampled at Lejre Historisk-Arkæologisk Forsøgscenter, Denmark. Three Butser Ancient Farm experimental sites were sampled, all in Hampshire, England, presenting a variety of tillage regimes: Butser Ancient Farm (spade and hoe tillage), the Butser Old Demonstration Area (ard and hoe tillage, left fallow to grassland) and the Little Butser site (ard, spade and hoe tillage, left fallow to grassland).

Except for the Silsoe soil bin, the experimental sites were chosen because they were locations where experiments in reconstructed 'prehistoric' agriculture were being or had been conducted. One of these, Butser Old Demonstration Area, was also one of Gebhardt's (1990) experimental sites, allowing the soil micromorphological assessment of short-term changes associated with leaving land fallow (untilled) after tillage with an ancient ard replica. Experiments at the soil bin were conducted to develop an understanding of the basic soil disturbance patterns created by an ard, in a situation in which soil variability is minimised and without the influence of other factors. The experimental sites are all sandy or silty loams, and are mostly types of brown earths; there was no attempt to explore implement impact variation on different soil types. Nevertheless, comparative archaeological samples show that the characteristics discussed in this project have relevance for a variety of soil types, because they are primarily defined by implement impact, and only then by secondary (pedogenetic) processes (Lewis 1998; 2002). Although experimental analogues are restricted in soil type, tillage regime and time depth, and thus cannot be directly compared to archaeological materials (Macphail *et al.* 1990a, 54), or even necessarily to each other, both experimental and archaeological results will be discussed in tandem in this volume.

The experiments focus mainly on one implement category – ards – although a large number of samples from tillage using other tools were taken from the Butser Ancient Farm experiments, where traction was not feasible for the duration of the research discussed here. Given the archaeological tillage evidence available from some areas, the apparent over-focus on traction implements may not be particularly skewed. Comparative samples belonging to other researchers from experimental tilling with ancient implement replicas and modern tools were also examined (see Lewis 1998, app. I), and have influenced the discussion in this volume.

It should be noted at this point, and will be brought up again throughout the volume, that this study was based on a three-year doctoral research project focused solely on soil micromorphology. Although in certain instances other basic soil characteristics were described (e.g. bulk density at the Silsoe soil bin – Lewis 1998), and some of these are generally discussed in places in the following chapters, at no time was a systematic study of other soil characteristics conducted for this research. This represents one of the limitations of this study. A second relates to certain soil factors, such as slope, soil type etc. Soils respond differently based on environmental factors and soil conditions; in this volume it has not been possible to discuss these variations in any satisfactory detail, given the focus on presenting the results of a limited study on brown earth soils.

Soils are defined here in a fairly standard manner: they are both physical bodies and systems comprising mineral and organic components, water and air, which develop on the surface of the earth through weathering and erosion of parent materials, addition of mineral and organic sediments, and the deposition and decomposition of organic matter. Soils develop horizons through the processes of soil formation (pedogenesis). The succession of horizons makes up the soil profile (Bridges 1978, 10-1; Williams & Smith 1989, 238). Horizons are described in this volume following the outlines given by Bridges (1978, 11-2) and Avery (1980).

Only those formation processes and soil types directly relevant to the experiments described here are discussed in detail. Richard Macphail rightly emphasises the fact that interpreting ancient tillage (or any land use) from modern soil profiles is based upon understanding what is anomalous for that particular soil (Macphail 2011 pers. comm.; see also Goldberg and Macphail 2006; Courty *et al.* 1989). However, certain actions of implements on the soil are held to have a similar immediate impact on soils of many types. For example, moving an ard through a rendzina or a podzol will still produce an ard mark with characteristic micro-organisational features, which are the main subject of this volume; the post-depositional survival and impact of these will certainly vary with local soil processes, and these must be studied and predicted with an understanding of each setting. As mentioned above, it has not been possible to discuss in any detail soil types and situations beyond those encountered in the experiments described here.

Disturbance features created by or related to the processes involved in tilling are frequently recognisable through the examination of macroscopic soil features and characteristics in the field. Many of these are also visible at a microscopic level, along with other characteristics which may be more difficult to gauge by naked eye or by chemical means, especially if one is concerned with the specific location, nature and relationships of micro-features within the soil profile.

Despite the premise that every macroscopically-identified feature should have a microscopic counterpart (and *vice versa*), relating microscopic characteristics to their macroscopic expressions can be problematic. First, there are certain characteristics which are not expressed so clearly at the different scales: for example, although it may be noted in the field that a particular soil horizon has a relatively high clay content, the nature of this clay (clean? illuviated? in the groundmass?) is not clear to the naked eye, nor discernible via chemical analyses (Courty *et al.* 1989, 24-5). Second, the question of how representative a thin section is of a soil horizon must be considered. Do certain microscopic characteristics that occur ubiquitously in thin section represent simply a zone of concentration which happened to be sampled, or are they a main profile characteristic? Detailed field recording of variations within horizons is vital for understanding whether features visible under the microscope are representative of a soil profile. This difficulty can also exist when moving between the various scales of microanalysis (Courty *et al.* 1989, 57), but may be addressed immediately during interpretation (the thin section is, after all, still available – the field profile frequently is not).

There are various scales at which soil micromorphology can inform about environment and culture. These have been defined as: the microscopic, the site-specific and local, the regional and the global (Courty *et al.* 1989). While the soil holds information useful to interpretation at all of these levels, the issue of comparison between scales must be considered (*ibid.*, 3-4; Macphail *et al.* 1990b, 168-70; Addiscott 1993). One thin section or one soil profile does not usually represent the entire soil body under consideration, although we generally draw overall conclusions based on one profile, or, on a larger scale, on the basis of a few profiles per site. The issue of selective, targeted or random sampling of a few profiles, as opposed to intensive 'blanket' coverage, is very important in discussing whether it is possible to *identify*, as opposed to simply *characterise*, ancient tillage in soils, as we shall see in later chapters (see also Carter & Davidson 1998; Macphail 1998a; Thompson *et al.* 1992).

As when dealing with features in the field, when examining soils using micromorphological techniques certain fundamental relationships are assumed to hold. First, it is assumed that soil physical and chemical characteristics may be related to specific pedological processes, notwithstanding the fact that it may not always be possible to identify which of a number of processes is responsible for creating certain features (Carter & Davidson 1998; Macphail 1998a). Second is the assumption of stratification following the geological law of superposition. It is possible to discuss changes in soil processes and activities that influence them over time through the concept of a hierarchy of features, in which not only are features seen to be created at various specific times relative to each other, but the interaction between earlier and later features in creating new characteristics is recognised (Kemp 1985; Mücher & Morozova 1983; Macphail *et al.* 1990a, 53-4; 1990b, 166; Gebhardt 1990, 31-2; Courty *et al.* 1996, 253), although only a relative temporal correlation can be proposed (Courty *et al.* 1996, 264).

In soil micromorphological analysis, as in field description, soils are described with regard to several inter-related aspects: microstructure (aggregate and pore organisation), groundmass (texture, organisation pattern and nature of the primary solid components (mineral and organic)), and pedofeatures (fabric inclusions, concentrations, depletions and other localised alterations) (Bullock *et al.* 1985; Courty *et al.* 1989, 64). Following the Bullock *et al.* (1985) descriptive criteria, which represent an internationally-accepted standard terminology, these main categories can be further broken down as follows:

- Microstructure: the size, shape, organisation and degree of development of soil aggregates (peds, clods, fragments), grains and pores in relation to each other (Bullock *et al.* 1985, 18),
- Groundmass: the characteristics of the principal mineral and organic components of the soil; their colours, types, shapes, sizes and size-grade ratio (texture), the arrangement of size grades in relation to each other (Bullock *et al.* 1985, 88-94).
- Pedofeatures: discrete zones which are distinguished from the groundmass based on origin, components and/or the ratios or arrangement of components (Bullock *et al.* 1985, 19). These are divided into: textural (mechanical concentration of a size grade of the principal components), depletion (loss of chemical components found in the groundmass), crystalline (*in situ* crystal growth in the soil), amorphous and cryptocrystalline (features composed of organic or mineral materials which are isotropic (black under crossed polarised light), e.g. iron and manganese nodules or impregnation), fabric (different organisation of the same principal components as in the groundmass) and excrement (soil fauna excremental features) (*ibid.*, 94-133). Pedofeatures are also described regarding all aspects listed above in microstructure and groundmass.

All thin sections discussed here were described under plane-polarised (PPL), cross-polarised (XPL) and oblique

Table 1 Experimental sites discussed in this volume

Site	Period	Tillage	Soil
Butser Ancient Farm, Hampshire, UK 0.58372°W, 50.56348°N	a) Reconstructed Iron Age b) Modern c) Modern	a) Hoe/spade/fork b) Pasture c) Background	a) Rendzina b) Rendzina c) Rendzina
Butser Ancient Farm Old Demonstration Area, Queen Elizabeth Country Park, Hampshire, UK 0.583382°W, 50.575735°N	Reconstructed Iron Age	Ard; grass fallow since 1986 (8 years before sampling)	Brown earth, calcareous colluviums
Little Butser, Queen Elizabeth Country Park, Hampshire, UK 0.977126°W, 50.982127°N	Reconstructed Iron Age	Ard/hoe/spade; fallow since 1982 (12 years before sampling)	Rendzina
Silsoe College Soil Bin, Cranfield University, Bedfordshire, UK 0.42378°W, 52.00852°N	Reconstructed Iron Age	Ard	(Fine) sandy loam; prepared soil
Lejre Historisk-Arkæologisk Forsøgscenter, Lejre, Denmark 11.5636°E, 55.3623°N	a) Modern b) Reconstructed Iron Age c) Reconstructed Iron Age	a) Pasture b) Ard; fallow since 1994 (1 year before sampling) c) Ard	Argillic brown earth

Table 2 Additional descriptive criteria for thin section summaries in this volume

Descriptive parameter	Decision
Texture	Texture is described after Bullock *et al.* (1985). All grains larger than 2,000μm are rock fragments. There is an additional division in some cases between coarse (20-50μm) and fine (2-20μm) silt.
C:f limit	A 100μm limit was used to describe coarse:fine ratios (coarse/very coarse/medium/fine sand = coarse; very fine sand/silt/clay = fine).
Amorphous organic components	Amorphous organics are not described beyond colour and general size. Organic components in general were not described in great detail for this study, which is biased towards structure, and textural and fabric pedofeatures. Some additional variation may exist in certain samples regarding manuring *vs.* non-manuring (e.g. at Butser Ancient Farm); see Lewis 1998.
Frequency (percentages)	All percentages are visual estimates of proportionate area, often in reference to an abundance chart (Bullock *et al.* 1985, 24-5). It was decided not to point count features for this project, and image analysis technology was not available at the time of the study.
Horizon distinctions	In some cases, horizons are named as context numbers given in the field ([204] etc.). In cases where the contexts ascribed in the field could not be distinguished in the thin sections, horizons and fabrics are described as 'Context 1' etc. In the experimental samples 'Horizon 1' etc. refers both to horizons *sensu strictu* and to subhorizons distinguished by some pedological characteristic.
Horizon names	Where the word 'litter' is used it refers to a surface horizon composed mainly of plant remains with cellular structure, apparently 'fresh' remains, with no further distinction as to origin or type. Ap/bAp refers to tilled horizons/buried tilled horizons. All other horizon names follow Avery (1980).
Calcareous pedofeatures	Following Courty *et al.* (1989, 175; after Folk 1959), micrite is <10μm, microsparite is 10-50μm, sparite is >50μm in size.

incident reflected light (RL) using a Wild Photomakroskop M400 and a Leitz LABORLUX® 12 POL S. They were described at a macroscopic level (by naked eye under transmitted light) in order to link field observations with thin section units, and at low (x20-x200) and high (>x200) magnifications (Courty *et al.* 1989, 70, 72, 75), after Bullock *et al.* (1985) and Fitzpatrick (1993) (and see Table 2). All percentages represent visual estimates of area. Full descriptions are presented in a standard descriptive format in Lewis (1998), and summarised in this monograph. Macroscopic pictures in this volume are mostly direct scans, although some were photographed in a lightbox. Microscopic images are scanned greyscale versions of photographs taken with the Wild MPS46 Photoautomat (Leica Heerbrugg AG 1991) and Wild M400 (Wild Heerbrugg Ltd. 1978). A colour image database is available in Lewis 1998.

Summary

Studying ancient arable land use through soil micromorphology involves identifying remnant indicators of the processes and activities involved in cultivation in thin section. Regarding ancient tillage, there are two major types of indicators which should be examined micromorphologically: profile or horizon characteristics associated with the impact of cultivation on the soil, and the characteristics of macroscopic tillage features themselves. Much primary research has focused on the former, although the latter may prove to be both the least ambiguous, and of the most use in relating microscopic indicators to macroscopic archaeological features. This volume discusses experimental study of both of these aspects, in comparison to archaeological remains, and presents a feature morphology-based approach to the study of ancient arable land use.

Chapter 2 The impact of ancient arable farming on soil profiles

Introduction

In defining prehistoric arable activity, Zvelebil (1994, 37, after Harris 1989) suggested that arable farming could be said to have been carried out where evidence of *systematic* soil tillage can be presented. Tillage is the systematic turning of the soil so that a seedbed is created for the production of crops. As such, it is not limited to arable farming of tillage crops (after Coppock 1964), and certain processes and activities described below are also applicable to some horticultural activities, as well as where cultivation is carried out for the production of grazing crops.

Different soils respond in different ways to cultivation, and the success of tillage and crop production depends both on specific soil properties and the strategies used by cultivators to deal with these. In locating prehistoric arable activities in the physical landscape, archaeological discussions frequently refer to soil or land fertility as a controlling, or at least prominent, factor in ancient decision-making. Land fertility and ease of tillage, based on soil-type distributions, have been linked to settlement location and technology at the time in question. Thus we have Neolithic peoples in Europe said to be settling 'lighter, better drained soils, such as gravel, sand, loess, chalk or limestone' (Clark 1952, 91), which were 'highly fertile (and) easily worked' (Whittle 1977, 242), and avoiding 'higher altitudes with their mountains and leached soils...as long as there were good low-altitude soils to cultivate and graze animals on' (Lindqvist 1987, 180). It has been assumed that people practising arable farming in any given region would have settled soils the most fertile for arable crop growth. Heavier soils, even if more fertile, were supposedly rejected in favour of those easier to cultivate. Soil properties and the technological ability to deal with them have thus been seen to be major limiting factors in settlement location for ancient societies with arable practices.

Soils exhibit varied characteristics based on a number of factors, collectively known as soil formation controls (Ellis & Mellor 1995, 97-113). These controls influence soil type variability, soil history, and soil potential for arable farming practices (Table 3), and they are discussed in depth in the pedological literature (e.g. A.S.A. 1982; Bridges 1978; Jenny 1941; Courty *et al.* 1989, 10-1; Schimel *et al.* 1985). Cultural practices such as tillage can have a major impact on these controls and their interactions. Soil development is divided into major process types: A and B horizon formation, homogenisation and ripening being a few examples. These are described in depth in textbooks on pedogenesis (e.g. Fitzpatrick 1971; 1986; Jenny 1941), and have been summarised regarding ancient soil development in England by Macphail (1987a). All of these processes have an impact on arable farming in both the short- and long-term. And they have an impact on the post-depositional environment of the features studied in this volume, which focuses primarily on the impact of implements on brown earths, and uses examples from a restricted range of environments and soil conditions.

Some predictions as to general ancient soil types at a location can be arrived at through mapping surficial geology (parent material), topography and hydrology (catenary position), vegetation and climate. But, due to microrelief and localised differences in all of the main soil formation controls, and because of variations in land use, soils may differ greatly over a small area, making it difficult to predict the specific nature of present-day profiles at any given spot before field examination; when the time factor is added it becomes even more difficult to interpret localised soil types without intrusive investigation. Arriving at a picture of prehistoric variability at or around an archaeological site entails detailed study of a variety of soil profiles, with a focus on their history and development (Rees 1979b, 3). And while this type of study produces a useful backdrop for description, soil types may ultimately have little or no bearing on whether or not arable farming is or was practised in any given location; even if appropriate technology is also in evidence archaeologically, prehistoric human choice of whether, where and how to farm is not immediately predictable from these indicators. Actual exploration of landscapes for ancient land-use indicators is the strongest interpretative strategy for archaeology.

Arable land-use practices and the creation of soil profile characteristics

In this volume the immediate influences of cultivation on the soil are considered to be primary processes. All processes occurring after this impact are deemed secondary, even though they may include primary soil formation processes. The impact of arable land use, and especially tillage processes, on soil qualities has been the focus of a great many studies in agronomy, geomorphology and pedology. The following sections draw upon this body of literature in describing the processes involved in cultivation with regard to resultant soil characteristics which archaeologists are able to identify. As noted in Chapter 1, this discussion is limited by the nature of the research presented in this volume, which is a soil micromorphological study of ancient tillage features, and does not aim to provide a complete, comprehensive description of all soil types or situations, nor of all approaches that can be used to study ancient land-use practices from soils; please refer to the many references cited for additional information on specific soil features and methods.

Cultivation processes: clearance

Although there are some arable systems which do not require complete clearance of pre-existing vegetation, such as tilling under or between trees, as has been suggested by some regarding Neolithic cultivation (e.g. Lüning & Meurers-Balke 1980; Meurers-Balke 1985), most arable farmers do clear land of forest, brush or grasses. Clearance is not an activity restricted to arable farming, and clearings may be created by many means and for many purposes, and need not necessarily be related to cultivation. Assuming, however, that clearance is normally a part of arable activity, we expect to frequently encounter evidence of clearance where arable farming was practised.

Archaeological site-specific clearance indicators include certain molluscs (e.g. *Pupilla muscorum)* and/or the impoverishment of the snail assemblage, soil pollen (e.g. of Graminae, *Plantago lanceolata*) and charcoal, as well as soil features such as tree bowls and the loss of the upper A horizon (Robinson & Wilson 1987; Tipping *et al.* 1994; Keith-Lucas 1986, 107). The interpretation of an anthropogenic origin for most of these characteristics must be argued on a site-by-site basis, and there are many cases in which alternate interpretations may be more appropriate (e.g. Langohr 1993; Exaltus & Miedema 1994).

The impact of clearance on soil varies with vegetation and soil type, as well as with method of clearance and factors such as topography and climate. There are three main site-specific types of impact which are often visible in archaeological deposits:

- loss or localised movement of upper organic soil horizons and supplies of organic matter (plants) through invasive mechanical disturbance such as digging, uprooting or ploughing (e.g. tree/bush bowls/throws, loss or local movement of upper A horizons through deturfing or 'sod busting'),
- additions of charcoal and ash through burning, or of degrading organic matter (if cut material was left on the surface to decay),
- secondary immediate and long-term physical and chemical alterations in the soil profile, related to degradation and diagenesis of organic matter, erosion by water and wind (both at a landscape-scale and localised), changing local hydrological regimes, the impacts of burning, specific chemical alterations in the soil, etc.

The uprooting and turning over of plants causes great physical disturbance to the soil horizons immediately involved, leading to mixing of horizons, and can result in the creation of archaeological features visible in the field (Courty *et al.* 1996, 263). The clearance of grasslands for arable farming by digging, for example, entails either the complete removal of upper horizons (deturfing), or the break up and sometimes inversion of the upper horizons (digging or ploughing in). The chopping down/cutting of plants will not necessarily result in directly visible macroscopic features in the soil, but can lead to many features associated with bare soil surfaces (discussed below).

Table 3 Soil formation controls (after Ellis & Mellor 1995; Oades 1993; Gebhardt 1990, 8; Courty et al. 1989, 10-1; Hassink et al. 1993)

Control	Main factors	Most influential aspects	Direct effects
Climate	Temperature, moisture, light, windiness	Precipitation – mean annual, seasonality and intensity; temperature – mean annual and seasonal fluctuation, vegetation type, evapotranspiration, growing season and rate	Additions of aeolian and water-based material, soil transport, rates of organic matter accumulation and decay, rates of mineral transformation, in-soil transfer of components, erosion, solution, weathering, translocation, drainage
Parent material	Chemistry, components, erodibility, permeability	Stone content, texture, mineral nutrients, acidity	Amount and type of mineral material available, rate of horizon development, soil weathering characteristics, drainage characteristics (transfers, chemistry), lateral losses
Biota	Amount, type, variety	Actions in, on and upon the soil, exposure to elements, temperature	Additions of organic matter, amount and type of litter, size and point of origin of organic material, transformation rates (decomposition, mineral weathering), transfer of organic and mineral matter, soil acidity, soil losses through gaseous exchange, surface and subsurface lateral removal, erosion rates, structural development and stability
Topography/ relief	Altitude, catenary position, aspect	Slope angle, slope position, temperature, exposure	Drainage, lateral movement (additions and losses)
Time	Succession, climatic change, progressive or regressive pedogenesis	Short-term (organics); long-term (geochemistry)	All of the above

Clearance may be accomplished using digging or tilling tools, and tilling is frequently carried out today to plough in the remains of previous crops or to break grassland. In Europe, 'heavy' ards and ploughs used with large teams of oxen are known historically for breaking virgin or long-fallow land, and Bronze Age rock art evidence of such large teams (4-6 oxen) also exists (Russell 1988, 39; Clark 1952, 100; Fowler 1971, 157). So-called 'rip ards' are proposed to have existed from the Neolithic period, with the same mode of practice envisioned as for these later 'heavy' ards. It has been suggested that widely-spaced ard furrows (e.g. 30cm apart) found archaeologically are not likely candidates for an interpretation of tillage for sowing, being better related to initial ground preparation activity (Fowler & Evans 1967; Thomas 1994). Hansen (1969) reports that clearing grass fallow with a replica prehistoric ard is very difficult, suggesting that hoeing or burning was carried out before tilling was done, or that grassland clearance for later cultivation was carried out using an ard with an arrow-shaped share. This type of clearance may be indistinguishable archaeologically from tillage for sowing if it is carried out in a systematic way over an area. Spades and shovels, along with hoes and other implements associated with cultivation may have been used for clearance in prehistory. This was one suggested use for proposed Neolithic hoes (Curwen & Hatt 1953, 64).

Clearance by fire can influence the soil immediately in two ways. There is an addition of charcoal and ash to the soil (Boyd 1982a; 1982b; Moore 1982; Goldberg and Macphail 2006), and there is the possibility of soil oxidation, especially where shorter surface vegetation is cleared (e.g. Korhonen 1985, 140-41). Using fire adds nutrients to the soil, enriching it for the short-term with base-rich ashes, and increasing the pH (reducing acidity and encouraging biological activity). Such burning only immediately influences the upper few centimetres of the soil (Gebhardt 1990, 15-6; Courty *et al.* 1989, 129; Dormaar *et al.* 1979, 80), but can have a long-term influence on soil chemistry. Burning is often used to enhance or maintain short-term soil fertility for arable farming, especially where soils are acidic (Ellis & Mellor 1995, 221). Burning as part of land-use strategies has, however, also been associated with acidification of soil (Gebhardt 1993), with a temporary decline in organic matter levels and crop yields (Dormaar *et al.* 1979, 80), and with prehistoric erosion (Bunney 1990). Experiments in 'traditional' slash-and-burn agriculture in Spain led to 'considerable soil degradation' in heathlands (Soto *et al.* 1995, 13); the long-term impact of burning varies greatly with specific soil conditions. Soil micromorphological study of slash-and-burn experiments at Umeå, Sweden, showed mixing of organic and mineral soil horizons and charcoal fragments as inclusions (Goldberg and Macphail 2006).

Charcoal inclusions in soils have been linked to clearance with fire in cultivation contexts in archaeology, sometimes as part of forest-, bush- or grass-fallow practices, with each sort of vegetation cover producing different amounts, types and sizes of ash and charcoal (Courty *et al.* 1989, 105-6). Soil characteristics such as leaching and loss of structure, and rare other inclusions, such as burnt stones found on 'field' surfaces (Evans 1971b, 52; 1971a, 1-18, after Dimbleby 1960), also provide evidence of localised burning. Burnt stone spreads have also been associated with tree clearance where found with tree-throw pits (Hey 1997, 109).

The study of the use of fire for field clearance on a site-specific basis requires integration of both regional and localised evidence addressing origin and degradation of charcoal inclusions in local soil contexts. Some lake sediment studies suggest that fossil charcoal, sedimentological and geochemical evidence for clearance, even when related to pollen evidence of vegetation decline, do not necessarily match records of localised fire activity, mainly reflecting fire activity in the larger catchment, along with related erosion episodes, and the pollen record is likewise not a site-specific indicator (MacDonald *et al.* 1991; Swain 1973; Cwynar 1978). Further means of distinguishing types of fires in the soil record have been discussed by Bellomo (1993). While natural fires are frequent in zones with coniferous/resinous forests, it is generally held that deciduous forests do not normally spontaneously burn (Iverson 1941; 1949; Gebhardt 1990, 18), and that ancient episodes of burning in the regional palaeoenvironmental records of much of temperate Europe therefore indicate the direct influence of human activities.

Microscopic charcoal in soils has been discussed by Patterson *et al.* (1987), Tipping (1996) and Gebhardt (2007), among others. Fragments >50μm in diameter are thought to be more representative of local fires than the smaller sizes, which could be deposited by other means, such as wind (Clark 1988a; 1988b; MacDonald *et al.* 1991, 65-8), but in soils charcoal becomes physically broken down over time, mainly through faunal activity (Courty & Federoff 1982). This natural process means that interpretation of whether or not a specific locality was cleared in the past using fire is difficult from this indicator alone. Midden deposits are known to have been dumped on ancient tilled fields, meaning that fire indicators such as charcoal and ash may show an amendment practice, or proximity to combustion features on settlements, but not necessarily plot clearance using fire.

In thin section studies, the presence of finely broken-up carbonised plant remains and mixing of these with the groundmass has been related to clearance, both on its own (Gebhardt 1993, 335-6), and associated with tillage and amendment (Macphail 1992a; 1989; Macphail *et al.* 1990a, 54-5; Mikkelsen & Langohr 1996, 148; Romans 1986, 128-9; Goldberg 1992, 159; Macphail & Goldberg 1990), or with slash-and-burn agriculture (Romans *et al.* 1973; Goldberg and Macphail 2006). Charred cereal grain is often found in buried soils (e.g. Macphail 1987c;

Tipping *et al.* 1994), and has been cited as a site-specific stubble burning or cultivation indicator, despite possible taphonomic issues.

Plant ash deposits and burnt soil have been characterised microscopically (e.g. Courty *et al.* 1989, 105-7, 109; Wattez & Courty 1987). Wood ash and fine charcoal found together in buried soils have sometimes been associated with amendment for arable farming (Romans 1986; Dockrill & Simpson 1994), or with burning episodes possibly related to slash-and-burn farming (Gebhardt 1993, 335-6), although the issue of fires not associated with farming must be addressed in such cases. The inclusion of charcoal and ash in the soil is said to favour clay dispersion, thus adding to the creation of textural features such as clay coatings with fine charcoal inclusions (Courty *et al.* 1989, 129; Mikkelsen & Langohr 1996, 148). Burnt soil fragments, in combination with fine and coarse charcoal have also been associated with slash-and-burn practices (Courty *et al.* 1996, 263). Microscopically, burnt soil fragments from midden materials found on possibly tilled land have been described as 'amorphous organomineral fragments which are distinctly red in oblique incident light' (Dockrill & Simpson 1994, 86), with this colouring being related to high iron levels and associated with oxidation through burning (Dockrill & Gator 1992).

Whatever the clearance method used, and whether or not land use is arable, the surface of the soil is exposed to erosion. Clearance has a great impact on the chemical and physical properties of a soil, influencing carbon, nitrogen and organic matter contents, as well as aggregate stability (Besnard *et al.* 1996). The devegetation of the uppermost horizons loosens the soil, and exposes it to raindrop impact (MacDonald *et al.* 1991, 55; Chartres & Mücher 1989; Courty *et al.* 1989, 131). Secondary alterations immediately following clearance include bare earth processes (surface crusting, sorting of surface materials by texture, leaching, in-soil transfers, downslope movement), and processes related to profile disturbance, especially where there are intrusive clearing practices (physical mixing of profile materials), and possible loss of nutrients. The impact of these processes varies with soil formation control factors (Table 3) specific to each location.

The impact of long-term clearance maintenance depends on the type and extent of vegetation cover. If a soil is left bare for extended periods of time (this can occur even if it is regularly cropped) it will normally see a reduction in organic matter content, because much of the plant matter is removed from the soil by clearance and harvesting, instead of being allowed to decay *in situ* (Ellis & Mellor 1995, 208). The rate of degradation of organic matter by biota changes with clearance of forest or pasture, and this can influence the rate of mineralisation (Nascimento *et al.* 1992). In addition, reduction in plant variety through monoculture and weeding leads to a reduction in the diversity of soil fauna. Even that organic matter which survives bioreduction relatively well normally is influenced by the physical break-up of the soil caused by tillage (Besnard *et al.* 1996, 495-6).

In the long term, clearance may encourage leaching, downslope movement and aeolian and/or alluvial transport, depending on the local environment. Soils left bare on slopes will erode, and colluvial deposits have been cited as evidence of erosion related to ancient clearance and cultivation (Gebhardt 1990, 3; Mikkelsen & Langohr 1996); this is discussed further below.

Cultivation processes: tilling and digging

Tillage has as its main agricultural purpose the creation of a tilth for crops grown from seed. The creation of tilth entails breaking down larger soil clods to create a finely structured seedbed (Ellis & Mellor 1995, 200), and increasing the amount of air by increasing pore space between soil aggregates. The creation of tilth enables better water flow, easier seed germination and emergence, easier root growth, and exposure of soil components to chemical erosion. Tilth is made by causing soil structural failure, and then by moving (mixing or turning) the 'failed' soil (Schafer & Johnson 1982, 17-8). The creation of good tilth depends on several factors, but mainly on the skill of the tillage operator and the soil moisture content. If a soil is too damp, the process of tilling may lead to compaction, which impedes drainage and root growth, and can increase susceptibility to erosion. On a finely-textured soil with poor drainage, however, increased macroporosity may improve the drainage situation (Ellis & Mellor 1995, 200, 202; Mackie-Dawson *et al.* 1989). Good tilth is especially needed for the broadcast sowing of seeds, but is routinely created today before drilling of the same, as well as prior to planting seedlings.

While digging often creates discrete features only, tilling produces both such features and a tilled topsoil horizon (Table 4), in which the upper soil horizons (L, F, H and Ah layers) are mixed with the A horizon, to produce a relatively homogenised layer (Ap) (Courty *et al.* 1989, 131). If the A horizon is relatively shallow, material from lower mineral layers (E, B or C horizons) or archaeological deposits can be brought up into the topsoil, altering its chemical and physical make-up. Buried soils which show mixing of horizons micromorphologically have been associated with ancient cultivation (e.g. French 1988; French & Pryor 1992). Mixing of layers can, however, be created by other means, including disturbance related to burrowing, tree throws, and human construction (Courty *et al.* 1989, 127, 140; Macphail & Goldberg 1990).

Tillage also moves soil laterally, mixing various parts of a tilled plot together, and forms part of the 'diffusive processes' influencing soil horizontal redistribution within and between fields (Govers *et al.* 1996, 929; Sibbesen *et al.* 1985). One mouldboard plough tillage event can move the entire ploughed horizon horizontally over at least 0.3m, in addition to inverting the profile (Govers *et al.* 1994, 469). This lateral intermixing does not directly influence horizons below the depth of tillage, and this is partly behind the strongly-expressed boundaries between plough zones and underlying layers (*ibid.* 476-7). Depending on how it is carried out, digging for planting or harvest does not necessarily involve major lateral soil redistribution, nor does it necessarily disturb the entire topsoil, only influencing individual points in an area. It does lead to horizon mixing within digging features, but does not necessarily create a homogenised Ap layer.

Basic structural characteristics of tilled soils

Soil structure describes how primary soil particles (sand, silt and clay) aggregate to form larger units (peds), and how these aggregates are arranged regarding void (pore) space (Bridges 1978, 20; Vogel *et al.* 1993, 301). The type of structure created by tilling varies with original soil structure and texture, type of implement (disturbance pattern), soil fauna and flora, and various secondary processes (Ringrose-Voase 1991, 785) (Table 5). Tillage causes a change in ped shape in the upper part of the soil through the break-up of aggregates (Dexter 1985), and this is related to porosity characteristics in the tilled horizon (Holden 1995). For instance, at Ashcombe Bottom, England, a Beaker-period site with two cultivation soils, the originally prismatic peds were seen to be broken up, presumably by implement action (Macphail 1990a, 266). Structure in tilled soils is relatively less influenced by biological activity than in untilled, due to lower biological activity because of a decrease in organic matter (where amendment is not carried out). This decrease in organic matter is also related to a reduction in aggregate size and a decrease in aggregate stability (Livingston *et al.* 1990; Oades 1993), and to an increase in erosion (Macphail 1990a).

Tillage can lead to angular or subangular blocky structure in topsoils with an original crack structure (e.g. Gebhardt 1990, 44-5), or a crumb or granular structure in originally massive or compact topsoils (Jongerius & Jager 1964, 495). As stated above, cultivation generally induces a reduction in biological activity, but when a soil is enriched through cultivation (e.g. through aeration, manuring), biological activity may also increase dramatically; this may produce a crumb structure. Jongerius (1983, 133) states a tilled soil can be recognised from this structure, if one knows it did not have such a structure originally. However, when the original structure is unknown, as in many archaeological situations, crumb structure is mainly taken to be indicative of biologically active topsoils (in particular mull horizons), and is also characteristic of many uncultivated soils.

Overall porosity usually decreases in the upper 20-25cm of a soil with tillage; while porosity between peds (interpedal) increases, porosity within aggregates (intrapedal) decreases, except in very clayey soils (Aguilar *et al.* 1990, 25; Jongerius 1970; 1983; Jongerius & Jager 1964, 495-6; Andreini & Steenhuis 1990, 86; de Olmedo Pujol 1983).

Tillage itself will not create an overall massive or apedal structure, although secondary processes related to disturbance, bare earth and flooding/irrigation can reduce or change the type of porosity to create a vughy or compact soil (Sandor & Eash 1991, 33; Ringrose-Voase 1991, 783-5; Jongerius & Jager 1964, 495; Thompson *et al.* 1990, 343; Courty *et al.* 1996, 263). The loosening through tillage leaves the upper horizon prone to leaching and secondary translocation and transport processes (discussed below), through disaggregation and dispersion related to the decomposition of organic bonding molecules (Reid *et al.* 1990, 200), and an overall change towards a massive structure is related to the decrease in stable aggregates in cultivated soils (Besnard *et al.* 1996, 495). Because of these secondary processes, modern tilled soils often have an apedal or poorly-developed angular blocky structure (Jongerius 1983, 113), but the impact of these processes depends greatly on local soil and environmental conditions.

Small subrounded aggregates with a relatively high organic component have been noted in tilled soils seeing regular manuring. These also occur rarely in non-manured and in uncultivated soils, and are suggested to be related to animal traffic (Gebhardt 1990, 45), although the shearing and rolling action of tillage implements could play a role in their formation (see Chapter 4).

Modern experiments show that the proportion of aggregates larger than 50μm in the soil decreases rapidly in the first several years after tilling begins, at which point aggregate destruction and creation rates become the same. It has been suggested that the lifetime of aggregates in a cultivated soil is only a few years (Besnard *et al.* 1996).

Table 4 The immediate impact of tillage and digging on soil horizon characteristics

Tillage	- creation of tilth; extensive soil disturbance to create seedbed, structural alteration of peds/porosity - local and plot-wide weed control, both immediate and long-term (influence on seedbank) - vertical mixing of horizons (topsoil, possibly subsoil); homogenisation, creation of Ap horizon - systematic lateral redistribution of soil materials across area of field or plot
Digging	- loosening of soil, structural alteration of peds/porosity in specific locations - vertical mixing of horizons and/or inversion within digging features and piles/banks - local and plot-wide weed control, both immediate and long-term (influence on seedbank)

All types of pores may occur in a tilled soil, with the overall porosity (amount and frequency of pore types) varying with the specifics of soil characteristics, tillage regime and secondary processes. Tillage has a dominant influence, with raindrop impact also being a strong factor in aggregate formation in tilled horizons (Martin *et al.* 1987, 222; Thompson *et al.* 1990, 330). In topsoil which has just been tilled, fissures (planar voids) are said to be the most frequently-encountered type of pore. Fissures are linear cracks created by the mechanical stresses involved in tillage, and are associated with blocky structures. Fissures are often infilled or compacted soon after tillage, and can form void space for plant rooting (Pagliai *et al.* 1984; 1989; Kooistra *et al.* 1985). Although planar voids have only occasionally been associated with prehistoric tilled soils (Fisher & Macphail 1985, 100; Macphail 1990a; 1992a), horizontal planes are known to develop in modern arable soils, often in apedal horizons, due to compaction, and their survival is related to a decrease in soil faunal activity (Martin *et al.* 1987, 222; Aguilar *et al.* 1990, 26; Jongerius 1983, 122). These pores also occur naturally in many lower soil horizons, where they may be created by shrink-swell processes (Macphail & Cruise 1996, 97).

The soil break-up caused by tillage can also create zones of packing pores (Ringrose-Voase 1991, 783), especially when deep tillage occurs, in which case large areas of compound packing voids may be found (Kooistra 1987). Under the influence of rainfall, flooding and/or compaction a vughy structure can develop (Pawluk 1980; Martin *et al.* 1987, 222; Thompson *et al.* 1990, 343). Other types of pores not directly caused by the mechanical break-up involved in tilling are also frequently encountered in tilled topsoils: channels (created by roots and fauna), chambers (created by fauna), and vughs and vesicles, where settling, compaction, crusting and freezing occur (Ringrose-Voase 1991, 783; van Vliet-Lanoë 1985a; 1985b). In buried soils, especially buried topsoils, vughy structures are often seen where other indicators suggest tillage (e.g. Macphail 1987c), but it may be difficult to rule out the influence of later compaction by burial. A spongy microstructure is associated with paddy fields and other flooded arable horizons (Gebhardt 1995, 34; for soil micromorphological indicators specific to paddy fields see Lee 2009; 2011). This type of structure has also been discussed in relation to plaggen soils (e.g. Bryant & Davidson 1996). Finally, a microstructure exhibiting collapsed voids has been related to irrigation for cultivation in arid zones, and the rapid drying of flooded surface horizons produces a vesicular porosity (Courty *et al.* 1989, 136-7).

Carter and Davidson (1998, 542) identify two main routes to structural alteration in agrarian soils:

- compaction related to the use of 'agricultural machinery' or animal traction (direct)
- changes related to declining soil stability and the amount and type of soil faunal activity (indirect – related to texture, organic matter levels, etc.).

They suggest that 'direct' microstructural characteristics of compaction are not to be expected from ancient cultivation methods (see also Macphail *et al.* 1990a), except possibly where animal trampling is involved, and that structural alterations related to tillage through 'indirect' means occur only under certain soil conditions (Carter & Davidson 1998). In Davidson and Carter (1998, 830), structural changes related to tillage are not included in a table of 'Predicted impacts of agriculture on soil micromorphology', which focuses instead on fabric and textural pedofeatures produced by various implements.

Non-mechanised tillage (especially in cases with no traction) has a relatively minor impact on soil compared to modern/historic 'heavy' methods. But this does not mean that it does not directly influence soil structure, or that it does not cause compaction sufficient to alter the long-term characteristics of soils. The creation of tilth implies structural alteration, as does the digging or drilling of holes for any purpose. The physical forces involved in the movement (tilling and digging) of any soil create direct structural change (disaggregation and aggregation) and zones of compaction and loosening (Gebhardt 1990; 1992). The shear forces involved when using implements such as ards or spades, mean that localised direct compaction, at least at the very edges (cuts) of dug features, should be expected to occur. This is discussed in subsequent chapters.

There are general associations between structural expression and tillage with various reconstructed ancient tillage implements. Gebhardt's research (1990; 1992; 1995) focused on structural characteristics produced by hoes, spades and ards, on the basis that pedality and porosity play a major role both in soil suitability for arable farming and in the influences of arable farming on soil profiles (Gebhardt 1992, 374).

Table 5 Characteristics of tilled horizon microstructures

Characteristic	Survival and identification potential (generalised; all are soil- and environment-specific)	Other possible generating factors
Crumb structure	Highly biologically active soil; survival of tilled structure is short-term. Where this can be shown to not have existed before tilling, it may be possible to infer increase in organic matter content	Crumb structure is created mainly by soil faunal action; not solely seen in tilled soils
Subangular blocky; fissures	Although many tilled soils have subangular blocky structure, showing good aggregation, so do many untilled – this structure cannot be directly related to tilling except where it is known not to be a structural type in the untilled soil	Rooting/shrink-swell, etc.
Angular clods in loosened uppermost part of horizon	Depends on how biologically active the soil is. Where these are subsoil aggregates there may be a better chance of survival	Rooting/other shearing forces
Small, rounded peds accumulating at base of tilled zone	'Rolling' – may have a decent survival rate based on the special characteristics of the base of tillage – see subsequent chapters; possibly only related to traction tilling	Must distinguish from colluvial rounded aggregates
Increase in interpedal porosity	Disaggregation of materials, resulting in much packing void space; distinguished in comparison to underlying/surrounding 'undisturbed' soil	Other disturbance factors
Decrease in intrapedal porosity	Very dense peds – especially in the context of increased interpedal void space	Other compacting forces
Decrease in structures created by biological activity (where non-manured)	Macropores; excremental features & fabrics; organic remains; need to be distinguished in a hierarchy of features to show relict biological features & demonstrate that their production is reduced; species dependent	Any factor changing organic matter type & amount, or soil chemistry – climatic, erosional, hydrological, etc.
Increase in structural characteristics tied to faunal activity (manured)	As above	As above
Ped fragmentation by implement action	'Anomalous' fragments which might be identified (e.g. crust fragments at depth)	Rooting, freezing and other shearing forces
Vughy structure (bare soil and/or in compacted zones)	Vughs are created by many factors – this structure would have to be clearly related to other features to be tied to tillage; survival rate probably low unless infilling occurs	Any sealing, internal slaking or compaction forces
Planar voids	Interpreted as shear planes related to tillage forces; survival and relationship to tillage (as opposed to other factors) may be low, unless associated with implement marks	Shrink-swell, roots, other shearing or compacting forces
Infilling of macropores with fine fraction	Illuviation is a normal soil process. 'Dusty/dirty' and silty clay infillings suggest disturbance-related. Silt/very fine sand infillings may relate to sealing, but may also be created by earthworm sorting (these are usually clearly distinguishable)	Any disturbance process
Creation of density boundary zones (loose/compacted)	Seen with implement mark features (inside/outside) and as horizon features, especially under mouldboard tillage; survival potential is good (based on macroscopic identification of such zones archaeologically), especially if secondary processes act to create horizon boundaries at the boundary of these zones	Any process creating loosened and compacted zones – including digging, burrowing – but the arrangement of these might be expected to differ from tilling zones
Collapsed voids	Survival potential seems good based on Courty *et al.* (1989), but will only be related to tilling in certain environments and systems	Flooding indicator, not necessarily linked to arable
Decreased porosity and macroporosity in B horizons	Density decrease in lower horizons related mainly to accumulation of translocated fine fraction; where these can accumulate and are not later disturbed, survival potential is high	Related to textural pedofeatures, which may be caused by other processes

Gebhardt's study of experimental tillage of brown earths with and without manuring, using various implements, demonstrates that the influence of low-impact implements can be very broadly compared to that seen under better-studied mechanised ploughing. Gebhardt (1990; 1992) found that cultivation loosens the soil in an explosive manner, reflecting fracture planes originally created by wet-dry processes, and that the size of resulting peds and depth of implement disturbance zone correspond to implement type. Loose and compact zones (implement impact zones) were observed, and textural features (plough pans, 'dusty' clay features – see below) were created using ancient implement replicas. Organic matter content was found to be equivalent in cultivated, manured fields and uncultivated grasslands (Gebhardt 1995, 29).

Much work on tilled microstructures has focused on macropores (>60μm r^2) and biopores (>500μm r^2) (McKeague *et al.* 1987; Andreini & Steenhuis 1990, 86). These large pores have a great impact on topsoil drainage, and any changes in these influence the erosion rates of the surface soil. Where there is no tillage, macropores run from the surface into the subsoil, enabling drainage. Conventional modern tillage, however, mixes the topsoil so that macropores may be destroyed or blocked; although tillage creates macropores itself, it results in a less continuous system, leading to a decrease in water infiltration and greater surface water run-off (Andreini & Steenhuis 1990, 86; Thompson *et al.* 1990, 330; Gibbs & Reid 1988; Pagliai & De Nobili, 1993; Edwards *et al.* 1988; Oades 1993).

Many macropores in the upper 20cm of a relatively non-acidic soil are created by earthworms, with the impact of these soil fauna varying with species and habitat (Diehl *et al.* 1995; Dexter 1978). Each year's tilling discourages earthworm activity in the upper horizons for a short time, even where manuring also occurs (Lee 1985, 282-6; Thompson *et al.* 1990, 348). Then earthworms return and macropores begin to increase in number and percentage volume. The direct impact of tilling with reconstructed 'ancient' implements on macropore formation has not been studied experimentally, but it is generally assumed that the changes associated with reduced incidence of macropores are not restricted to soils tilled with modern machinery. However, the apparently short-term nature of this impact (Andreini & Steenhuis 1990, 86-100) suggests that any such changes inferred from ancient soils (e.g. Macphail 1990a, 265-6) should be interpreted with caution.

Earthworm-based structure (e.g. crumb peds, excremental fabrics) is expected to dominate in certain upper soil horizons, such as those on calcareous parent materials under grasslands (Atkinson 1957, 220; Shipitalo & Protz 1989). Where such structure is expected and not found, a decline in soil fertility (organic matter content) and/or increase in acidity, often related to evidence of clearance and cultivation, has been cited as the reason for decreased faunal activity (Macphail 1990a, 265-6; Macphail *et al.* 1987; Cornwall 1963; Atkinson 1957, 220-1). Experimental buried soils at Overton Down, England, however, showed a general loss of organic matter simply through burial, and increased acidity was also found where burial occurred under a turf stack (Macphail 1994b, 22). An increase in acidity was suggested from the loss of earthworm burrows through the activities of later acid-tolerant microfauna in buried old surface layers at the sites of Carn Brea and Chysauster, England (Macphail 1987c; Macphail 1990a). But at Ruguellou, France, thin section analysis showed earthworm activity continuing despite the presence of indicators of acid-tolerant microfauna (Gebhardt 1993, 336). Predicting ancient rates of earthworm activity, while a logical indicator of variations possibly related to cultivation, is a very complicated approach. In addition, the continuation or, in some cases, onset of earthworm presence in certain ancient soils has been explained through probable addition of manure to the soil or liming (Atkinson 1957, 221; Courty *et al.* 1996, 263; Gebhardt 1993, 336; Macphail 1987c; Dockrill & Simpson 1994).

Manuring can offset tillage impact on structural change, as the related increase in organic matter content leads to more stable aggregates, and encourages faunal activity (Bouma 1969). Gebhardt (1990, 45) found that aggregate size increased in tilled plots with yearly manuring, and manuring has been associated with the creation of a more open porosity (Courty *et al.* 1996, 263). In some cases, faunal activity may be so high that structural changes related to tillage may be negligible in impact on the overall structure (Gebhardt 1990, 34; Borchert 1964). Experimental study of manuring in podzols suggests that an increase in faunal activity creates homogenisation of horizons even in acidic soils (Goldberg and Macphail 2006, 265).

Davidson *et al.* (1992, 57) report that at Castle of Wardhouse, Scotland, they were able to estimate an 18-month time delay between final cultivation and burial of a tilled topsoil through examining the rate of faunal reworking. This was represented by excrements and microaggregate fusion seen in thin sections from both a buried soil and overlying recently-tilled topsoil. The results of experimental earthwork studies and archaeological comparatives (Macphail & Goldberg 1995; Courty *et al.* 1996, 262; Maltby & Caseldine 1982; Macphail 1987c), suggest, however, that faunal activity does not cease immediately after soil burial; if this is the case it is hard to see how subsequent rates of reworking can be predicted.

The presence of features created by earthworms, such as vermiforms and calcitic granules, cannot be seen as necessarily indicative of cultivation, but their absence in soils which are predicted to contain them could suggest recent tilling, or, where evidence for more acid-loving microfauna appears, an increase in soil acidity (Gebhardt 1993, 336). Other soil fauna which produce excremental features include enchytraeids, which have normally been linked to coniferous forest, moorland and tundra soils, but are known to be abundant in both arable and pasture soils with high and low pH levels (Dawood & Fitzpatrick

1993). Mite pellets are also commonly seen in tilled topsoils, usually in root or other organic remains (see Fitzpatrick 1993, 135-41 and Courty *et al.* 1989, 142-6 for more about soil fauna).

Structural changes may also occur in horizons under the zone of tilling, through the pressure of implement movement and any traction team or machinery, and with the secondary movement (translocation) of fine soil components down the profile. Tillage leads to the creation of structural boundaries in the soil. Large aggregates and clods remain in the surficial layer, which has a loosened structure (increased interpedal porosity) (Dexter 1976; 1979; Macphail *et al.* 1990a). Small aggregates and fine fraction components accumulate at the base of the tillage zone, and this, coupled with implement pressure on the surrounding soil, results in reduced porosity. That features akin to modern plough pans (Collins & Larney 1987; Kooistra 1987; Kooistra *et al.* 1984) can be produced using ards was shown in experiments at Hambacher Forst, Germany, where a 1cm-thick silt 'pan' developed at the base of the implement disturbance zone (Gebhardt 1990; 1992, 377; Macphail *et al.* 1990a). Such pans have been seen archaeologically; Fisher and Macphail (1985, 98), for example, describe a 'low porosity plough pan with fine charcoal' related to ard marks. The compaction zone at the base of tillage impact is exacerbated where puddling also occurs (Jongerius 1970; 1983). The creation of this zone may lead to secondary characteristics because water infiltration may be reduced, leading to internal slaking (*ibid.*) in soils susceptible to such processes.

Structural changes associated with textural properties can occur at depth in A or B horizons in tilled profiles. While the upper Ap may become more porous due to loss of fine fraction through translocation through rainfall impact, the location in the profile where these fine materials accumulate (if they do) becomes less porous due to infilling of void spaces (de Olmedo Pujol 1983). B horizons in cultivated soils have shown lower porosity, especially less macroporosity, and a lower level of bioactivity than those under forests (Thompson *et al.* 1990, 329; Aguilar *et al.* 1990, 26). In soils with a substantial clay content the decrease in macroporosity may be related to the deposition of leached fine fraction material from the A horizon under tillage impact (Jongerius 1970).

General groundmass and fabric characteristics of tilled soils

Soil texture may be greatly influenced by tillage processes. Coarse components can be added to the upper part of the profile through physical interaction with lower horizons, including parent material. Fine components may be released through clod break-up and increased physical and chemical erosion, and fine materials may be moved (thus changing textural characteristics in specific parts of the profile, as noted above), or completely lost to the profile (Courty *et al.* 1996, 263). In addition, fine materials become more susceptible to dispersion and movement under surface puddling (bare earth processes) and water flowing through the profile (translocation) (Thompson 1983, 537; Greenland 1977). Under the influence of water, the bonds involved in flocculation (attractive forces keeping the particles together) can become reduced. This leads to the breakdown of soil aggregates and the separation of the basic textural constituents of the soil. Dispersion increases the effectiveness of other disaggregation processes. The most important soil properties regarding disaggregation are related to the bonding properties of organic matter, iron, aluminium oxides and oxyhydroxides, and the exchangeable sodium percentage of the soil. Break-up of clods depends on cohesion between clay particles and between the aggregates themselves through surface tension; soil moisture content has a great impact on both of these factors (Le Bissonnais 1996, 427-8; Kandeler & Murer 1993; Spoor 1969a). Bare earth processes and translocation are discussed below.

Aggregates from lower horizons may be pulled up into the topsoil with tillage, and upper horizon material taken down, even with relatively shallow tillage if the topsoil is a thin horizon. Features of horizon mixing related to tillage have been found to occur experimentally (Gebhardt 1990, 45), and archaeologically (Macphail 1987c; French and Marsh 1999; French 2003), and have been related to both cultivation and colluviation (Macphail 1990b, 189-91). Gebhardt (1990; 1995) describes 10cm long 'comma shaped' fragments of B horizon material with sharp boundaries in an Ap horizon at Malguénac, France. These are related to ploughing, and were noted in the field. Where mixing includes cultural layers it may be more easily identified (e.g. mixing of midden deposits – Macphail 1987b). At Tanners Hall, England, a 'dark earth' horizon was interpreted as cultivated based on the presence of two different soil fabrics mixed together and evidence for organic matter incorporation (Macphail 1983, 249). And mixed dumped debris with A horizon material led French (1988, 347) to interpret amendment and cultivation at Borough Fen, England. As mentioned previously, however, horizon mixing may have many origins, such as tree throws, up-rooting as part of land use, or bioturbation (Courty *et al.* 1996, 263).

Tillage mixes the soil and creates a horizon which may be comprised of many materials and soil fabrics, organic matter and textural features (Courty *et al.* 1989, 131). The relative lack of fine earth components, especially clay, in tilled surface horizons has been related to the processes of slaking, runoff and translocation. In extreme situations, the remaining surface material may be coarse and 'densely packed, (with) washed grains whose surfaces are not coated by fine particles' (*ibid.* 155). Through the direct influence of tillage and these secondary processes an 'agric' horizon may be created. With strong translocation of fine components, the entire groundmass may become enriched with clay, silt and associated fine organic particles. This type of horizon is said to be created by long-term cultivation of a soil, and is characterised by a group of textural features: grain coatings, void infillings and pans (Courty *et al.* 1989, 131). These pedofeatures are discussed further below. In situations where overland flooding or irrigation occurs, an increase in the proportion of very fine particles (clay and silt) may be seen, and this changes the textural and drainage properties of the soil (Courty *et al.* 1996, 263).

Tillage mixes the upper profile, which normally consists of litter layers, biologically-worked topsoil, and underlying more mineral topsoil layers. This mixing is often expressed through the presence of fabric pedofeatures, but may be strong enough to result in a horizon which is dominated by mixed groundmass fabrics (e.g. Lewis 2003). With repeated tillage these may become a homogeneous fabric, frequently with abundant textural and fabric pedofeatures (Courty *et al.* 1989, 132; Macphail 1987c). Mixing of soil horizons in this way has been related to all types of tillage implements (Davidson & Carter 1998; Macphail 1990a; 1990b), and is also seen within individual implement mark features (Lewis 1998).

The type of distribution fabric a soil has is dependent mainly on the proportion of clay, silt and sand, and the location of these size grades in relation to each other. Tilled soils frequently exhibit a poorly-sorted matrix, with a porphyric distribution pattern (Aguilar *et al.* 1990, 26). In some topsoils with a high silt and clay content, the action of secondary processes such as illuviation and puddling may be so strong that after the physical disruption of tillage an almost complete separation of fine and coarse components occurs, creating a groundmass composed of a mosaic of clayey and sandy zones (Jongerius & Jager 1964, 495).

Each soil has an orientation fabric, which is the pattern of organisation of zones of oriented clay particles. Tilled soils often exhibit stipple-speckled, crystallitic (Aguilar *et al.* 1990, 26) or granostriated fabrics. At Potterne, England, where tilled soil horizons overlay material with a very low clay content, an increase in clay levels and speckled birefringent fabric in the tilled horizons was attributed to increased weathering of subsoil material related to cultivation (Macphail 1987b).

Clay in the groundmass becomes oriented due to shear stresses and plastic deformation, and through deposition on pore and particle surfaces. Strongly-oriented fabrics are thought to have undergone more shrink/swell (wetting/drying), shearing and smearing processes than weakly-oriented fabrics. Tillage, as a major type of disruption, has been shown to increase the incidence of strongly-developed orientation fabrics, through smearing by machinery (impeding drainage; common at the base of certain modern plough layers), and after wet cultivation (leading to massive structure and poor drainage). Other possible causes of such fabrics include shrink/swell processes involved in wet-dry and freeze-thaw cycles (Ringrose-Voase 1991, 787-8; Miedema & van Oort 1990).

Smearing is a term used to describe the influence of stresses associated with compaction under the pressure of machine weight and motion. Smearing fabrics are strongly-oriented (unistrial), and may occur at the edges or bases of tilled features or at the base of cultivation layers (Blackwell *et al.* 1991; Ringrose-Voase 1991, 787). It is unclear whether these are created by lower-impact tillage technologies.

The impact of tillage on inclusions

There may be an increase in the number of rock fragments in a tilled soil, and this may affect the chemistry of the topsoil (e.g. Courty *et al.* 1989, 7). Tilling and translocation of fine particles down-profile may result in a net upward movement in the profile of all coarse inclusions, and a surface accumulation of inclusions such as stones (Govers *et al.* 1994, 477; 1996, 943), creating features such as stone pavements where freeze-thaw or winnowing processes are also active (Washburn 1979; Johnson & Hansen 1974). Lateral and vertical movement of inclusions (such as artefacts) in tilled soils has been discussed by Odell and Cowan (1987), Boismier (1995), Yorston *et al.* (1990), Schofield (1990), Yorston (1990), Roper (1976) and Reynolds (1982; 1987; 1988), among others.

It has been suggested that homogenised buried A horizons which are stone-free are created by earthworm sorting under grassland, as all items too large for worms to move tend to accumulate at the base of the topsoil (Cornwall 1953, 131-2; Atkinson, 1957; Evans 1971b, 34; Ashbee *et al.* 1979, 282; Jones 1986, 55; Romans 1986, 125), and some authors have connected this to lack of cultivation (e.g. Hey 1997, 109). However, Romans (1986, 185) found a 5-7cm deep, relatively stone-free surface layer interpreted as relating to earthworm activity in a soil at Walls, Scotland, which was known to have previously been ploughed, but then turned over to grazing. In addition, some of what have been identified in the field as stone-free horizons have, upon micromorphological analysis, turned out not to represent simple worm-sorted topsoil materials – being reinterpreted at Strawberry Hill, England, for example, as chalk slurries, said to represent colluvial material from intensively cultivated rendzinas uphill (Allen 1992). On the other side of this issue, ancient topsoils with stone inclusions cannot be taken on this basis alone to be tilled, as many other forces can lead to the inclusion of stones in soil profiles.

Although tilling with metallic cutting blades may create angular rock fragments, this is not necessarily the case with implements such as ards, with the exception of instances in which a tilling implement cuts into an underlying solid parent material which is relatively soft, such as chalk. In cases where ongoing, especially traction, tillage occurs, stone angularity should be reduced overall, due to the rolling motion both of the inclusions themselves and of soil aggregates along them (Wood & Whittington 1960, 334). It may be, however,

Table 6 Some secondary processes following tillage (after Govers et al. 1996; Conry et al. 1996; Courty et al. 1989; Nettleton et al. 1987; Bridges 1978; Jongerius 1983; Le Bissonais 1996: Gebhardt 1990; Harlan 1995, 218)

Process	**Morphology/macrofeatures**	**Creation (all are soil- and environment-dependent)**
Dispersion	Involved in creation of slaking and translocation features	Physical break-up of aggregates; separation of basic textural components through water action
Raindrop impact breakdown	Microaggregates; involved in dispersion and creation of micro-features	Physical displacement of fragments or basic textural components under impact of rain
Surface slaking	Crusts (compacted zones); vesicles; sorting by texture; creation of microaggregates; translocation; features of re-deposition (infillings/coatings) down-profile	Structural collapse when saturated; lateral and vertical translocation of separated basic textural components; redeposition as sorted deposits
Internal slaking	Sorting by texture; laminated sorted fines infillings/coatings; traffic sole or plough pan (compacted zones); translocation; features of redeposition (infillings and/or coatings) down-profile	Rising groundwater influencing compacted layer at depth in the soil leads to loosening, sorting and downwards translocation of sorted basic textural components; redeposition as sorted (layered) deposits; plough pan associated with downward pressure under modern heavy equipment, not thought to develop under low impact tillage
Shrink-swell	Structural alterations; disaggregation into smaller macroaggregates and microaggregates	When certain clays wet and dry they swell and shrink, creating shear forces in the horizon
Leaching	Horizon showing loss of soluble salts and bases, e.g. decalcified	Soluble salts and bases are taken up into soil solution and removed from original location or from the profile through downward percolation of water
Translocation	Eluvial and illuvial horizons; many microscopic features	Movement of primary textural components, plus organic matter and acids in solution or suspension from one horizon (eluviation); redeposition lower in the profile (illuviation)
Podzolisation	Eluvial and illuvial horizons; iron pans; heath formation; acidification of profile	Combination of leaching, eluviation of organic acids, organic matter, Fe, Al, Si etc.; illuviation of these in lower horizons; increased acidification based on changes in litter form
Erosion (off-site)	Soil loss; colluvium/aeolian deposits elsewhere	Slope processes and rainfall lead to soil movement; loosened and dry soil is susceptible to wind erosion
Erosion (on-site)	Infilling, reduction of relief; inclusions of nearby 'foreign' material	Lateral redistribution of soil material through tillage, raindrop impact, soil creep

that tillage using spades and hoes which interact with certain parent materials, such as chalk, creates angular rock fragments. The impact of these tools appears to have seen little discussion in this regard. Inclusions of components related to other arable practices (e.g. manuring) are discussed below.

Secondary profile and horizon processes associated with tillage

The secondary processes occurring in tilled soils produce many features which can be seen macroscopically, although several other characteristics are only defined in archaeological soils after micromorphological study. The creation and preservation of features related to these processes are soil- and environment-dependent; in certain types of soils, such as those discussed in this volume, the features created by these processes form some of the most important potential indicators of ancient tillage known (Table 6), and they will be described in some detail in succeeding chapters. Most of these are based on the break-down of soil aggregates which comes about through a combination of the mechanical action of tilling, the influence of rainfall, and the loss of structure associated with lowered organic matter levels and microbial processes.

Leaching, translocation, eluviation, illuviation and podzolisation

Leaching, eluviation and illuviation are fundamental soil profile formation processes (Courty *et al.* 1989, 8), but their impact on tilled horizons is here considered to be secondary to the primary processes of tillage itself. Leaching is a process by which soluble salts and the bases Ca, K and Mg are taken up into the soil solution and removed from the soil through the action of downward percolating water. The loss of bases leads to acidification of the affected horizon. In some cases the removal of substances from an upper (eluviated) horizon is followed by their redeposition at a lower (illuviated) level in the soil profile. Usually these translocated materials consist of finely dispersed humus (highly degraded organic matter), clay particles and weathering products, such as iron, which accumulate in B horizons. With increasing acidity through leaching, types of surface litter and Ah horizons change, such that rainwater can take soluble substances (organic acids, colloidal organic matter) from the upper horizons through the soil profile.

The solution causes leaching and breakdown of clay mineral structure, releasing Fe, Al and Si etc., leaving a skeleton of grains (e.g. in a bleached Ea horizon) (Bridges 1978, 30-1). The release of iron makes humic molecules insoluble and they bond to clay molecules, creating an organo-mineral complex (Gebhardt 1990, 10). Lower in the profile iron oxides, Al and organic matter may be deposited in B horizons. This is the process of podzolisation. Agricultural activity has been said to be one trigger of or intensifier of podzolisation, although this is a long-term process, reliant on local soil formation factors (Bridges 1978, 30-1).

Translocation of clay occurs where dispersible clay is available in an upper horizon, the soil sees downward percolation of water, a lower horizon which can receive the clay is present (or clay will not flocculate and will be lost to the profile), and the moved clay remains intact (Gebhardt 1990, 10; Nettleton *et al.* 1987, 38). Mechanisms for clay movement and redeposition have been summarised by McKeague (1983) (see also Duchaufour 1982; Wieder & Yaalon 1978).

Coatings and infillings of sorted or intermixed clay, silt and sand particles are formed by the deposition (illuviation) of materials which have been eluviated from upper layers. Perhaps the most widely-cited pedofeatures associated with tillage in archaeological studies are agricutans and 'dusty', 'dirty' or silty clay coatings and infillings (Fig. 1; Table 7) (e.g. Macphail *et al.* 1987; Macphail 1986b; 1987a; 1987b; 1987c; Jongerius 1970). Agricutans are laminated textural features associated with disaggregation and internal slaking. They consist of 'fine-grained plasma, very fine sand grains and very fine organic particles' (Courty *et al.* 1989, 131). Found in soils with a relatively silty texture, they comprise layers of particles sorted by texture (separation of sand, silt and clay) and redeposited in lower horizons. Agricutans are said to develop below the tillage zone, and are limited in their development by the factors which control eluviation/illuviation (Jongerius 1983, 119-20; Courty *et al.* 1989, 131, 155).

These features may be contrasted with papules, which are inclusions of clean limpid clay (Fig. 1), often interpreted as representing leaching under forest (e.g. Gebhardt 1993, 336). Limpid clay features have been associated with slowly-moving water, while 'dusty' clay coatings are related to quick flooding and movement of coarser particles (Gebhardt 1991a, 88). The underlying principle is that if the soil is not mechanically disturbed and disaggregated, the silt-sized mineral and organic particles found in agricutans and 'dusty' clay coatings should not be taken into suspension (Jongerius 1970; 1983; Macphail *et al.* 1990a, 56; Bullock *et al.* 1985, 109-15; Gebhardt 1990, 33).

In soils with significant clay content 'dusty' or 'dirty' clay coatings and infillings may develop, and possibly silt and sand infillings may form in the surface horizons and at depth in the profile. These deposits are associated with dispersion of topsoil material under soil puddling or flooding, and with mechanical disturbance followed by translocation of fine components down-profile (Greenland 1977; Thompson 1983). They are primarily composed of clay which is 'dusty' with intermixed silt-sized opaque particles (organic and mineral) and non-opaque grains, and often with sand inclusions. The fact that these textural features and agricutans are often deposited, and can survive, at depth is very important for the potential identification of disturbance in archaeological contexts, as the accumulation of translocated materials below surface layers may lessen the chances of their being disturbed by subsequent activity. In addition, where accumulation of fine materials is strong lower in the profile, decreased permeability at this point may have a great influence on subsequent profile characteristics (Greenland 1977).

Although these features are a major characteristic of tilled horizons, their creation and survival is restricted to certain types of soils and certain conditions. Agricutans, for instance, are said to be commonly found in cultivated sandy loam or silty soils (Courty *et al.* 1989, 155; Jongerius 1983, 119-20). Soils with low water stability and high clay content are very strongly influenced by tillage, and many clay-humus and sand textural pedofeatures result (Jongerius & Jager 1964, 495). In acidic horizons with a low clay component, '...the chemical alterations caused by tillage are not strong enough to change the soil solution so as to produce movement of fine soil particles (and) we should not expect to find prominent textural features related to cultivation in these soils' (Jongerius 1983, 133; see also Macphail *et al.* 1987). Rendzina soils have been said to show little fine particle translocation due to their calcareous nature, a high amount of organic matter and related biological disturbance – these soils are thus relatively resistant to textural pedofeature formation, and those pedofeatures that do form are quickly disrupted through bioturbation (Macphail 1993; 1994b; Whittle *et al.* 1993; Gebhardt 1992). These are the types of soils often found as buried soils in the Wessex region of England, and they are focal to discussions of ancient land use and landscape in this important region of prehistoric Britain (see review in French *et al.* 2007), and form a core part of this study (see below).

While 'dusty' clay coatings have been demonstrated to occur under certain conditions with tillage (Jongerius 1983), they have also been found in many disturbed soils under bare surfaces, regardless of whether these have been cultivated or not. For instance, raindrop splash on bare earth where vegetation has been removed by fire, making it susceptible to leaching, has been seen to create such features (Boulbin 1976; Limbrey 1975; Gebhardt 1990, 33; 1993, 334-5; Macphail & Goldberg 1990; Macphail *et al.* 1990a, 54-5; Mikkelsen & Langohr 1996). They have also been related to tree throw disturbance (Macphail 1987d; Macphail *et al.* 1990a; Macphail & Goldberg 1990; Gebhardt 1993, 336), and slope movement in both soils and sediments (Courty *et al.* 1989; Goldberg 1992, 150-3, 163). Irrigation and flooding can produce an abundance of clayey silt intercalations and coatings, although the combination of these with other features such as vesicles and certain types of crust formation may distinguish these processes from tilling (Courty *et al.* 1996, 263; 1989, 137). Gebhardt states that these textural features mainly indicate bare soil conditions and that '(la) mise en culture devra donc être confirmée par d'autres méthodes' (1990, 33). These features cannot be linked to a sole cause (tillage) even in modern studies, and ancient examples must be interpreted cautiously.

Figure 1 Examples of clay-dominated textural pedofeatures

Where soil constituents and climate are favourable to clay translocation (Nettleton *et al.* 1987, 38), physical disturbance such as that created by tillage should result in bond and aggregate disruption which should create 'dusty' or 'dirty' clay textural features. Such soils should possess 'dusty' clay coatings when disturbed; if such a soil does not possess these features, it may seem reasonable to suggest that it has not been disturbed in any major way (and therefore not tilled). However, there is a chance that some textural features are lost or masked through biotic activity or leaching (e.g. Macphail 1989; Romans & Robertson 1983b, 64). It is also important to note that the creation of such features depends on whether or not a soil has been decalcified (Duchaufour 1982). Gebhardt (1995, 34), for example, suggests that 'dusty' clay coatings were not produced during her experiments in tillage because the soils were not sufficiently decalcified.

In addition to being mixed in with clay in the textural features discussed above, silt infillings and coatings may form (Macphail 1992b; Macphail *et al.* 1990a). Silt is translocated in suspension, often mixed with organic matter. A silt-dominated pan was created during ard tilling experiments at Hambacher Forst (Gebhardt 1990). Silt:clay panning features consist of fine silt grades (<20μm) mixed with almost equal part clay (Lewis 1998). Coarse silt pedofeatures are also created in tilled fields, especially in regard to crusting and to macropore infilling (Fig. 2). Fine and very fine sand-sized components can also form features in tilled soils, mainly as part of depositional crusts or pans in which strong separation occurs (see below). Sand and coarse silt separation is also associated with earthworm activity, found in both tilled and untilled soils.

Romans and Robertson (1983a; 1983b) interpreted the implement of cultivation at several archaeological sites in Scotland on the basis of the presence of textural coatings and infillings. On the principle that such features may accumulate at the depth at which the boundary between loosened (tilled) and denser/compacted (untilled) soil occurs, an accumulation zone of textural features may represent the base or depth of a tilled layer. In their study, depth of tillage from the surface was suggested to be marked by the depth of relatively high frequencies of oriented clay coatings. This was linked to cultivation implements, with a highest frequency of these pedofeatures at 4-5cm depth being related to hoe cultivation, while highest frequencies at depths of 12-14cm were said to represent ard ploughing (*ibid.*; Macphail 1987a; Macphail *et al.* 1987). The latter figure fits Gebhardt's (1990) experimental results. Thus, at sites where both other indicators of tillage and a clear ancient surface (from which to measure) are present, it is said to be possible to use the depth of such pedofeatures to infer type of implement. Following this line of reasoning, Macphail (1987c) suggested that an accumulation of 'dusty' clay coatings in the upper few centimetres of a buried soil under a kerbstone at Chysauster, England, represented possible hoe cultivation, and that cultivation indicators occurring throughout the upper buried soil at Ashcombe Bottom, England, could be related to ard tillage (Macphail 1992a). Macphail proposed a 6cm depth for ard tillage at this site (shallower than stated above, possibly related to truncation of the buried soil), and that the tilling indicators found underneath this depth represented 'phases of accumulation (occurring) at the same time as cultivation' (*ibid.*).

Clay pedofeatures may be created by many disturbance and other processes in soils (McKeague 1983; Bullock & Murphy 1979), especially when earthworks overlie them. The location of ancient topsoil surfaces is variable, as the upper horizons of buried soils are often lost, unrecognisable, or it is impossible to measure original topsoil thickness due to compaction, truncation and post-depositional transformation of the A horizon (French 1990, 308; 1985-6a; 1985-6b; French & Pryor 1993; Jewell & Dimbleby 1966; Ashbee *et al.* 1986, 23; Crowther *et al.* 1996, 111-2). Romans and Robertson's

(1983a; 1983b) proposed hoe-related features at 4-5cm depth were in cultivation ridges found under a barrow mound, while those related to ard tillage at 12-14cm depth were located under a henge and a rampart. It is not clear from the published material how the impact of compaction under earthworks in these two different situations was determined. Little is known about tillage boundary zones with regard to hoes, with some indications that at least certain types of hoes may till as deeply as do some ards (see Chapter 3), and there has been no work on the creation of various density zones in ridges, where 'hoe-tillage' microfeatures have been interpreted, although Romans and Robertson (1983b, 63) suggest that accumulation occurred at the pre-ridge land surface.

The conditions needed to use textural feature depth to infer not only depth of tilled zone, but also implement type, will be very rarely met in the archaeological record. However, the base of tilled horizons and the cuts of tillage features do constitute zones for accumulation of fine materials, and the investigation of these locations in the profile, and of the microfeatures found in them, has great potential regarding the characterisation of ancient tillage. Comparisons between archaeological and experimental evidence for implement disturbance zone depth are discussed in the following chapters.

'Dusty' clay and silt infillings and coatings (sometimes in the groundmass) and agricutans located in B horizons have also been associated with tillage disturbance. Thompson *et al.* (1990, 347) discuss possible agricutans consisting of layered silt and clay, presumably created through slaking of both size grades through the soil (after Kwaad & Mücher 1979; Pawluk 1980), although their layering is similar to that seen in surface crusts. Macphail (1987a, 343) suggests that 'dusty' clay features in the buried Bt horizon underlying Kilham long barrow, England, represent clearance and agriculture. But 'dirty' clay coatings seen in the Bt horizon at the North Level sites in England (French 1985-6b, 44) were associated with clearance only.

Tilled soils are suggested to have more clay coatings in general than do grassland or forest soils, due to their greater disturbance and lower degree of faunal activity (regarding reworking of features) (Thompson *et al.* 1990, 347). This observation may be of limited value to archaeological interpretation unless an adjacent contemporaneous untilled ancient soil can be found with which to compare a cultivated horizon. The most that can be accomplished without such comparatives is a history of the development of coatings in the soil.

Hydromorphic leaching can lead to manganese and iron pan formation in situations where strong density and/or chemical boundaries exist in the profile. Such boundaries may be created at the base of tilled layers, and may be identified in the field (Courty *et al.* 1989, 153-4). Panning can produce waterlogging above the pan, and this may be an important mechanism in the onset of peat formation in appropriate environments (but see Mitchell 1972 and Conry *et al.* 1996). Iron pan formation is not created by tilling itself, nor is it only created in relation to tillage (*ibid.*, 283), but it may sometimes be used to identify the lower boundary of a tilled horizon. Coatings and infillings composed of amorphous organic matter, fine silt and clay, and often organic or sesquioxide-stained are frequently seen in podzolic sands, and sometimes their development may be associated with ancient tilling episodes (e.g. Bjerre Site 7, Denmark; see below and Lewis 1998).

Romans and Robertson (1983b; also see Gebhardt 1990, 34) suggest that a sequence relating depletion haloes or bleached rims around stones in the soil to profile evolution from brown forest soil through to peaty podzol can be related to deforestation, grazing and arable activity, especially in comparison with evidence of earthworm activity levels in thin section. Stones in brown forest soils are meant to show no weathering crusts, while those in acid brown soils have iron-stained rims, possibly with some bleached weathering crusts. Podzolised soils show bleached rims over 1mm in width, with the widest rims occurring within a few centimetres of 'mineral ground surface', and rim thickness decreasing with depth (Romans 1986, 126-7). These features do not, in themselves, constitute evidence of land-use practices, but in comparison to other profile developments (e.g. earthworm activity, evidence for stone clearance, other lines of clearance and tillage evidence, etc.) they have been used to reconstruct the history of acidification of cleared, tilled and pasture land in regions with a cool, wet climate (*ibid.*, Romans & Robertson 1983b; Dockrill & Simpson 1994).

Decalcification of calcitic soils has been related to bare surface processes, leaching, erosion and mixing; where seen with other disturbance indicators this has been related to ancient cultivation (Limbrey 1992, 55-7).

Raindrop impact, slaking, sealing and crust formation

Raindrop impact can cause small pieces or elementary particles of aggregates to break off and be displaced. This impact is greater on a wet soil, because the aggregates are weaker to begin with. A greater sand content is meant to produce a greater amount of material disturbed with raindrop splash. Vegetation cover protects against raindrop impact (Le Bissonnais 1996, 427; Verhaegen 1984, 504).

Slaking is the collapse of structure in the upper horizon(s) of a soil upon water saturation. When aggregates become wet, air within them is compressed, vughs are created, and aggregates can become broken down into microaggregates and basic textural components. The influence of slaking decreases as the volume of trapped air decreases, and when clay content increases. Slaking leads to a compacted zone through changes in pore shape, the dispersion of aggregates and infilling of pores with fine particles (Le Bissonais 1996, 427).

Surface sealing is caused by disaggregation and dispersion followed by the translocation of fine particles in the surface layer and reduction in porosity due to infilling of surface voids with fine particles. Although soils that are prone to crusting may have higher erosion rates, soil losses actually decrease as surface sealing increases. Crusts form on bare soil surfaces and within surface-connecting voids in the uppermost horizon (Jongerius 1983, 118; Le Bissonais 1996, 427; Norton 1987, 127-8; Verhaegen 1984, 505; Bresson & Boiffin 1990).

Tillage is not a necessary factor in crust formation, but the disaggregation created by tilling does favour crusting in soils that are prone to these processes. In addition to bare soil and disturbance, animal trampling has been seen to play a role in crust and lamina formation, both in fields (Mikkelsen & Langohr 1996, 147; Beckmann & Smith 1974) and in buildings with earthen floors (e.g. Macphail 1989). In agricultural soils, surface slaking and crusting often have a colluvial component, with fine fraction materials both puddling at the surface, and moving down the slope of ridges, cultivation hills and furrow sides, creating slaking deposits that may be archaeologically identifiable (Currie 1994, 69; Courty *et al.* 1989, 132; Goldberg 1992, 161; Macphail *et al.* 1990a, 61-5).

Boiffin and Bresson (1987) identify two stages in surface crust formation from small aggregates: disaggregation and fine fraction deposition around surface aggregates, which leads to partial and then continuous surface seals. Partial surface seals are first developed in those areas of the surface with the highest fine component concentration, with additions from structural collapse of larger clods, associated with infilling of macropores with fine materials. The surface layer becomes compacted when soil is bare, leading to a decreased infiltration rate (*ibid.*; Valentin & Ruiz Figueroa 1987, 404; Valentin & Bresson 1992, 225; Evans & Buol 1968). The initial sealing is defined as a structural crust (with vesicles), created by the impact of water drops, with the formation of a depositional/sedimentary crust, formed by particle translocation, following once sealing is established (McIntyre 1958; Tarchitsky *et al.* 1984; Valentin & Bresson 1992, after Chen *et al.* 1980).

A depositional crust is formed by particles moving in suspension and being sorted by size during surface runoff. These have layers sorted according to texture, and are deposited so that thick crusts occur in microdepressions and thin layers are seen in higher areas (Valentin & Ruiz Figueroa 1987, 404; Boiffin & Bresson 1987). The crust layers often consist of silt and fine sand grains alternating with silty clay layers (which may be poorly-oriented and 'dusty' with organics) (Bresson & Boiffin 1990, 320-2; Courty *et al.* 1989, 155). The porosity in these features is often vesicular, with a fine crack structure appearing upon drying when clay content is sufficient (*ibid.*). Continuous surface seals result from continuing disaggregation and redeposition of fine components, to the point that lower areas (microdepressions) become infilled (Boiffin & Bresson 1987). Micromorphologically, surface crusts in arable fields have been seen to form several patterns (Valentin & Ruiz Figueroa 1987, 404-5; Evans & Buol 1968; Pagliai & La Marca 1979).

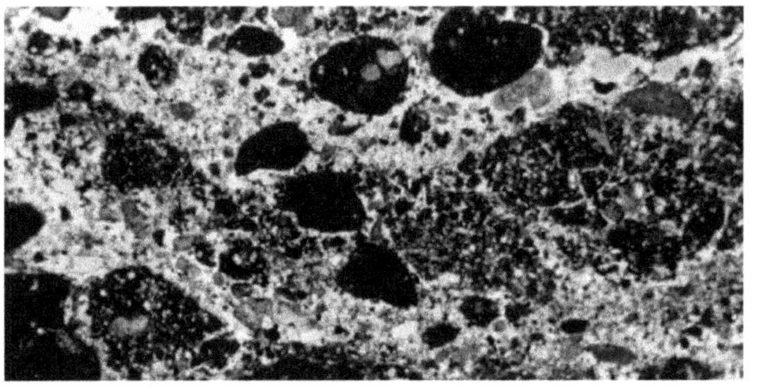

Figure 2 An example of coarse silt- and fine sand-dominated textural pedofeatures (Butser Ancient Farm PR19/1, PPL) 1 mm ⟷

In situ surface crusts are not frequently found in archaeological examples of agricultural soils, even where a clear upper topsoil horizon is seen and the soil is prone to the creation of such features. This is presumably related to their relatively short lifespan, as they are quickly reworked by bioturbation and altered by further surface processes (Courty *et al.* 1989, 155), being found usually only where relatively rapid burial has occurred (Currie 1994). Buried apedal surface crusts related to ancient burning and/or cultivation have been identified micromorphologically at some archaeological sites (Macphail 1987c; 1992a; Romans & Robertson 1983a; 1983b). Broken-up crust fragments found at depth have been interpreted as representing tilling and/or irrigation or flooding (Courty *et al.* 1996, 263; 1989, 137).

Internal slaking has been described in modern cultivation studies. It occurs under the influence of rising groundwater meeting a compacted layer (such as a furrow base). Soil particles become separated by size grade, and are deposited in size-sorted layers. Near the base of the tilled layer, internal slaking can lead to a change in structure, with the zone becoming further compacted (traffic sole/plough pan), influencing water infiltration rates. The depth of this compacted zone varies with depth of working (Gebhardt 1990, 34; Jongerius 1983, 119, 122; Beckmann & Smith 1974).

Table 7 Microscopic structural and textural features characteristic of tilled horizons (after Ringrose-Voase 1991, 787; Jongerius 1970, 321; 1983, 119, 122; Courty et al. 1989, 125; Le Bissonais 1996, 425-7; Norton 1987, 128, 134; Aguilar et al. 1990, 25-6; Gebhardt 1990; Macphail et al. 1990a; Macphail 1992a; Kooistra 1987; Bouma 1969; Kwaad & Mücher 1979, 183; Courty 1996, 4-5)

Process	**Description of features**	**Creation**	**Related processes**
Smearing	Strongly oriented fabrics (unistrial – Blackwell *et al.* 1991) located at edge/base of tilled layer/features	Pressure from weight and motion of machinery	Can impede drainage
Crusting/ surface slaking/ compaction	Horizontally continuous dense surface zone of lower total porosity than underlying horizon; 0.1-1cm thick, can be laminated	Disruption of surface aggregate structure by raindrop impact (reduction in aggregate size), detachment of fine aggregates and particles; decrease in frequency and size of packing voids *vs.* underlying (smaller aggregates more closely packed); 'explosion of enclosed air'	Impedes downward drainage, encourages surface water runoff; clearance of vegetation
Macropore infilling and percentage decrease	Fine fraction infilling macropores; loss of macropores seen up-profile relating to change in soil fauna associated with increased acidity	Infilling through physical disruption of tillage, plus translocation of fine fraction; increased acidity	Drainage; podzolisation
Aggregate break-up	Soil aggregates are smaller in tilled *vs.* no-till soils; 'broken' larger peds	Cutting action of machinery; sometimes increased faunal activity; microaggregates created as a result of surface slaking	Drainage; illuvial pedofeatures
Compaction	Plough pan/traffic sole; fine fraction infilling voids at base of plough zone, plus physical compaction	Weight of machinery, illuviation of fine materials/puddling, decrease in activity of soil fauna and microorganisms. At depth under mechanised ploughing, zones of compound packing voids and loosely packed aggregates can be quickly compacted into a plough pan (in 3 yrs) – general compaction of soil fabric – especially active with deep tillage, influencing sandy loams and brown earths at *c.* 25cm; can be offset by ensuring organic content and biological activity remain high	Impeded drainage may result; internal slaking of fines; related to decreased activity of soil fauna & microorganisms; reduction in organic matter; presence of earlier plough pans/compacted layers; groundwater rise
Internal slaking	Fine-grained B horizons (clay, silt & very fine sand and organic particles); compact, speckled layer; agricutans	When reduced downward flow of water; slaking of overlying horizon; downward movement of colloidal material (illuviation). Change in structure due to water saturation	May occur due to compaction or surface slaking; reduces rooting/drainage; organic matter content important (reduces)
Puddling	Vesicular porosity; horizontally-oriented fabric; 'dusty' clay coatings in pores	Vesicles form under puddled surface crust, later filled with fine fraction as water drains	May occur due to compaction

The impact of time of tilling

Tilling often occurs in the spring and/or later summer/autumn. The advantages of tilling in the spring and autumn are based on the water-retention and organic properties of the soil – at these times it is normally easier to obtain the best degree of aggregation and of disaggregation for seedbed preparation. Under the action of light rainfall, disaggregation is encouraged, but if the soil is too wet structural damage may occur (Brady 1974, 58-9). As with all of the above features, processes vary with soil conditions and environment, as well as crop.

Time of tilling has a great impact on the features created in the soil. The time at which the soil is left bare, and thus susceptible to erosion, is very important in regard to the creation and survival of features such as surface crusts. In Britain, for example, spring tilling is followed by a season of great biological activity in the soil, and this may destroy crusts and produce faunally-reworked structures. Autumn tilling leaves the soil open to crusting over the following wet season, and may produce more substantial secondary features than would be seen with spring tillage (see Chapter 5).

Soil redistribution

Lateral and downslope soil movement under tillage and/or colluviation may be interpreted in soil micromorphology from inclusions of soil fabrics originating from outside the local profile ('foreign' inclusions). Soil on slopes can become very dense (and laminated) due to the addition of disturbed A horizon material from upslope (Jongerius 1983, 131, citing Borchert 1961). Courty *et al.* (1989, 133) suggest that tilling of an upslope area where soils are relatively thin, followed by vertical mixing as the implement interacts with lower horizon materials, can be identified in topsoil horizons downslope through inclusions of rounded fragments of colluviated B (or C) horizon material. Transported rounded fragments were seen at Potterne, England, associated with ancient cultivated horizons, and related to colluviation (Macphail 1987b). Macphail (1987c) also mentions 'foreign' subsoil fragments in one thin section from Chysauster, England; these were related to colluvial and cultivation activity, and were associated with a possible lynchet. In addition there are fragmented crusts and coatings related to surface and internal slaking found in the lynchets from this site (Macphail *et al.* 1990a, 63). Romans (1986, 128-29) describes micromorphological indicators related to amendment with ash (charred organic matter, charcoal, peat fragments and related diatoms) within colluvial material at Scord of Brouster, Scotland. A different type of inclusion was seen at Potterne, England, where midden deposits saw addition of 'foreign' topsoil material which, along with many other features, has been related to cultivation activity on the midden (Macphail 1987b).

The processes of soil redistribution (tillage, soil creep and raindrop erosion) can lead to changes in soil bulk density, in-field soil movement, and in-field erosion (soil banks, infilling and reduction of relief, etc.). Bulk densities may be higher in the A horizons of cultivated fields than in uncultivated areas, although manuring (and associated increases in macropores) may offset this (Sandor & Eash 1991, 33-4).

Off-field erosion is downslope overland flow under the influence of water, and aeolian erosion (Govers *et al.* 1996); presumably, tectonic events leading to mass movement could also be included here. Alluvial deposits have also been related to agriculture, in that they may reflect erosion from bare earth or disturbed areas upstream (Pryor 1995, 102). The impact of erosion related to soil flooding and run-off varies greatly with texture, as well as slope and other local factors. For instance, a greater sand content is meant to produce a greater amount of material disturbed with raindrop splash and therefore with overland flow and downslope run-off. The impact of high silt fractions is variable, with some studies suggesting that a high silt content increases stability, while others finding that the silt fraction increases run-off (Verhaegen 1984, 504). Soil erosion and weathering processes have been discussed on a geomorphological level by Gerrard (1992), and these have been applied to the Holocene by Thornes (1987) and van Vliet-Lanoë *et al.* (1992), among others.

Erosion history in relation to agricultural history has been a popular area of research (e.g. Bunney 1990; Davidson 1980; Runnels 1995; Olson 1981; Dearing *et al.* 1990; Lagerås & Sandgren 1994; Verill & Tipping 2010). Erosion evidence is complex, and suggests that the timing and location of soil loss and profile changes vary greatly, even in small regions (for example, compare Allen 1992, 1994, Evans 1971a, Fisher 1982, Thomas 1996, and Bell 1981b, 1982, 1983 and 1992 regarding the chalk downlands of southern England). Soil erosion can be caused by many means, and often occurs through a combination of factors. Land-use practices, such as the onset or intensification of arable farming or grazing, are only one set of possible causes (Imeson *et al.* 1980, 32-3). Colluvial and wind-blown deposits may relate to ancient cultivation practices, but they alone do not always allow one to pinpoint either the exact location or date of ancient arable farming (Evans *et al.* 1984, 28).

Investigations by French *et al.* (2007) found variable results from an erosion study of a chalk downland valley in southern England, with a noted lack of deposition of eroded sediments during most early agricultural periods, but definite colluvial and aeolian deposition at later periods. Dated colluvial and alluvial deposits have been related to many periods in the region (see French *et al.* 2007 for a review and summary), varying between valleys and within each valley, and showing the complexities of trying to directly relate eroded deposits to broad histories of agricultural impact. For a detailed discussion of local erosion and vegetation changes that highlights how complicated the interpretation of anthropogenic associations with such evidence can be see Bennett *et al.* (1992).

Downslope erosion is responsible for the creation of colluvial deposits, and partially for the loss of soil to river systems and the sea (Imeson & Vis 1984, 593). Vegetation has a great impact on downslope erosion of the uppermost levels of the soil. Terwilliger (1990, 553-4) explains that a rich vegetation cover helps to prevent downslope soil slippage by reducing the amount of water held in the soil, such that a higher intensity and length of rainfall may be sustained before the soil reaches its capacity. However, intense storms may cause downslope erosion regardless of vegetation cover, and productive vegetation may increase the chance of creep and slippage by increasing the permeability and water-holding capacity of soils. The addition of organic matter through practices such as manuring can aid in reducing this effect through improving aggregation properties (Ellis & Mellor 1995, 208).

Cultivation processes: fertilisation/amendment/ 'anthropogenic' soils

Manuring compensates for losses of fertility associated with cultivation. Soils which are manured may contain as much or more organic matter than their uncultivated counterparts. Manuring creates conditions in which aggregation is stronger, and it is thus held to improve soil structure, although this is not always the case (Gebhardt 1990, 16; after Bouché 1979). Curwen and Hatt (1953) suggest that the initial establishment of permanent fields required an understanding of the importance of manure; this is suggested to occur in the Bronze Age in northern Europe (Courty *et al.* 1989, 7; Simpson *et al.* 1998; Bakels 1988). For instance, Nielsen (1984, 160) proposes that the long-term use of field systems on very 'poor' soils indicates that manuring was practised in late Bronze Age-early Iron Age Denmark, and Englemark (1992) suggests farmyard manure was used in the late Bronze Age in Sweden. Experimental studies on podzols in Sweden have also concluded that manuring must have been required for successful cereal crop production on such soils in prehistory (Macphail *et al.* 1998; Englemark & Linderholm 1996). However, the establishment of fields and field systems need not be related to arable farming, and there is no reason to suggest that prehistoric arable farming was focused on successful crop production over the long term, nor that all ancient arable farming was successful (Lewis 2008; 2010). In England manuring and liming are most often associated with the onset of mouldboard ploughing in Roman times or later (Macphail *et al.* 1987; Fowler & Evans 1967), although some chemical and soil micromorphological evidence suggests an earlier date (e.g. Macphail *et al.* 1990a; Macphail 1992a, 210). Manuring is normally identified in field archaeology through pottery and other refuse spreads (Wilkinson 1982; 1989; Mercer 1970; 1981b), particularly if these are found in prehistoric field systems or in relationship with prehistoric farmsteads (Barker 1985, 52).

The addition of manure leads to changes in structure related to stronger properties of aggregation and the impact of bioturbation by soil fauna, along with an increased retention of organic matter (Pitty 1978; Courty *et al.* 1989). In a tilled and manured field, organic matter is usually relatively quickly integrated with the mineral groundmass, although sometimes fragments of manure are still identifiable by their fabric. Reduction of organic matter content and change in size of aggregates related to tilling without manuring have been associated with an increased level of erosion (Livingston *et al.* 1990). Tilling and erosion together are said to lead to a tendency over time for subsurface material to be moved into the upper horizons, and this has a great impact on the amount and decomposition rate of organic matter in the profile (Schimel *et al.* 1985; Bouwman 1989; Mann 1985, 241).

'Foreign' fabrics have often been related to tillage by proxy, through association with amendment activities. For instance, peat fragment inclusions in 'traditionally' cultivated mineral topsoils have been associated with additions of hearth materials (where carbonised peat is seen) and turf (where uncarbonised peat is seen) as manures (Davidson & Carter 1998, 833; Adderley *et al.* 2006). These additions have also been suggested through the presence of relatively deep topsoil and soil textural characteristics; through analysis of organic manures documentary evidence of manuring planticrues with fresh turf has been supported (Davidson & Carter 1998, 833-34, 836-7).

With manuring, tilled soils have as much organic matter as grassland soils, while unmanured tilled soils generally have much less (Gebhardt 1990, 44; Courty *et al.* 1996, 263). Comparison of micromorphological indicators with macro- and microfossils, and the results of chemical analyses, is an important means of studying possible manuring and rotations. Chemical analyses of buried soils can show organic matter level and phosphate enrichment (inferred manuring) or poverty (inferred tillage with or without manuring), both in comparison to predicted levels based on modern studies, and between soil horizons and across a site area (e.g. Provan 1971; 1973; Prøsch-Danielsen & Simonsen 1988; Liversage *et al.* 1987; Macphail *et al.* 1987; Romans & Robertson 1983a; Quine 1995; Dockrill & Simpson 1994; Englemark & Linderholm 1996). It is difficult to predict what organic matter levels would have been in ancient soils, as the main part of the soil involved – the topsoil – is known today to be continually changing, and 'broad variations are seen in cropped zones, depending on the type of farming practised' (Eidt 1984, 29).

Research combining some or all of phosphate, organic matter content, magnetic susceptibility, lipid, stable carbon isotope, phytolith and soil micromorphological analyses has led to the characterisation of manuring components in prehistoric podzolic and possibly plaggen soils (e.g. Dockrill & Simpson 1994; Dockrill *et al.* 1994; Simpson *et al.* 1998; Englemark & Linderholm 1996; Englemark 1992; Macphail *et al.* 2004). It may be difficult to distinguish organic components directly related to arable activity from those which may occur through pastoral activities or even naturally, except in specific instances, e.g. the peat manures discussed by Davidson & Carter (1998). Much archaeological work has focused on biomarkers, distinguishing components related to manuring and manure sources. It is suggested that any carbonised or non-organic components of manure, being more likely to survive, have more to offer in regard to identifying ancient cultivation than other organic components (Carter & Davidson 1998, 542), but the results of other methods, such as lipid analysis (e.g. Evershed *et al.* 1997; Simpson *et al.* 1998), suggest that certain organic components might survive and can be used to identify the species of origin of manures, although the use of chemical methods to identify manuring has been criticised regarding interpreted source and dating of materials (Thomas 1990, viii; Bethell & Mátė 1989). Research into magnetic susceptibility, enhanced phosphate and biological molecules suggests that while such markers may show potential in this regard when combined with other methods, they can be alternately interpreted as relating to burning, occupation deposits, and grazing/penning (Evershed *et al.* 1997).

Soil elemental compositions have frequently been used to distinguish activity areas (e.g. Linderholm & Lundberg 1994; Asplund *et al.* 1989; Middleton & Price 1996; Ullrich 2010), and have also been applied to land-use interpretation of ancient arable fields. For example, Entwistle *et al.* (1998) report that low P enrichment in the historic fields they studied may relate to loss of crop material at harvesting (see also Edwards 1983), and possible distinctions were seen in the same study in Ca and Sr between fields amended with shell sand and those not amended.

In addition to indicators of rates of soil faunal activity, discussed previously, manuring may be evinced in thin section by the presence of inclusions such as charcoal, ash, shells, bones, phytoliths, pottery sherds and organic-rich soil fragments. Spherulites and coprolitic inclusions may also be present. The latter may be characterised by microfabric, food inclusions and structural characteristics to animal type. These are also frequently found in pasture soils (Courty *et al.* 1991, 145; 1989, 113-5, 118; 1996, 256-7, 262). Spherulites are found from many types of sites where stabling or penning has occurred, including caves as well as open-air sites, and are often found in burnt dung materials (Canti 1997; 1998; Brochier 1993; Brochier *et al.* 1992; Macphail 1998b). Frequently related to manuring practices, grain coatings and void infillings composed of mixed amorphous organic matter, charcoal, phytoliths, plant fragments and mineral particles have been found at sites such as Phoenix Wharf, England, dating to the Bronze Age (Macphail 1990a, 266).

Regarding other sorts of amendment, tilling into an underlying calcareous subsoil horizon can have the effect of liming the topsoil. The actual practice of liming is said to date to Roman times in Europe, and is mostly associated with the historic period in areas such as Britain (Limbrey 1975; Courty *et al.* 1989, 7; Goudie 1981). In the Wallonie region of Belgium, however, evidence for liming suggested to date to final La Tène comes from the site of Arbre, and possible identification of Roman liming practice has been presented from Bruyelle (Fechner *et al.* 1998, 81; Fechner 2000).

Additions of organic matter, manure and other amending materials, in combination with the increased soil volume (often temporary) and overland movement of soil associated with cultivation frequently leads to the thickening of a topsoil horizon. For example, cultivated horizons which may represent anthrosols are seen overlying building destruction layers at some archaeological sites (e.g. Macphail 1992b, 274), and in some areas sufficient topsoil thickness for cultivation was achieved only after a period of breaking up and extensively amending soils with manure and seaweed (Crummy *et al.* 1992, 33).

The most visible change in prehistorically cultivated and abandoned soils in Sandor's (1992) study in New Mexico was increased A horizon thickness, attributed to the local redeposition and cultivation of topsoil within terraces. This was also seen in the comparative region of Peru's Colca valley, with terracing leading to A horizon thickening, through a combination of hand movement of soil plus sediment transport (*ibid.* 227, 237). The hand movement of alluvial silts to amend a 32 hectare interpreted anthropic soil in Méidote, Colombia, leading to its having a continuous thickness of *c.* 1m, is suggested to indicate the efforts of a large group of people, with organisation of several villages under a central leadership (Santiago Mora *et al.* 1991, 81).

Soil thickness is not, however, always necessarily a good indicator of tillage practice. Variables affecting thickness also relate to environmental factors (e.g. slope, wind erosion, litter accumulation) (Lerche 1994, 141; Gerrard 1992). Other factors, such as truncation and compaction when buried, may sometimes limit the usefulness of thickness as a land-use indicator for ancient topsoils.

Plaggens are 'artificial' soils created by the dumping of manuring and waste materials, turf and/or peat and/or seaweed, often mixed with other sediments. These are ploughed in to create a homogeneous, often dark grey topsoil horizon, which becomes very thick (up to 1m) and phosphate-enriched (Pape 1970; van de Westeringh 1988; Courty *et al.* 1989, 134; Davidson & Simpson 1984, 76; Bryant & Davidson 1996, 813; Bakels 1988; Conry 1971; 1974). Although many plaggens are thought to be quite young in date (Conry & Mitchell 1971; Shiel & Askew 1988; Spek 1992), Bronze Age plaggen-like soils are known from northern Scotland, being composed of turf, seaweed and manure; some are said to have been cultivated intensively for barley production (e.g. Dockrill and Simpson 1994; Simpson 1985; 1993; 1994; 1995; 1997; Simpson *et al.* 1998; see also Bell 1981a re. seaweed).

The term 'dark earth' has been used to describe 'dark coloured European anthrosols formed during post-occupation abandonment' (Macphail *et al.* 1998), mainly regarding historic urban soils. These soils are dark grey-black and consist of mixed soil material and occupation debris. Macphail (1983, 245-6) distinguishes two types: one developed from refuse disposal, which is rich in organic matter and associated with cess, the other developed from dumped soil material, possibly for local cultivation. These soils have been studied micromorphologically (e.g. Macphail 1981; 1983; 1994a; 1994b; Cammas *et al.* 1995; 1996). Plaggen soils, dark earths and anthrosols are seeing a great deal of recent study worldwide (e.g. Macphail *et al.* 2003; Cummins 2007; Arroyo-Kalin 2008; Goodman-Elgar 2008), and the reader is referred to these sources for updated information.

Cultivation processes: the effects of sowing, weeding and harvesting on the soil

Broadcast sowing is commonly related to traction agriculture (and especially harrowing – Martin Jones, 1997, pers. comm.), while drilling or planting is often associated with hoes, sticks and spades (Harlan 1995, 225). Different crops may be associated with different tillage regimes. For example, the Romans are said to have sown wheat after three or four tillings, while barley was sown after two (Russell 1988, 120).

Regarding site-specific indicators of arable farming and tillage, soil pollen of cereal-type plants has been seen in thin sections from buried soil horizons, although it is suggested that only when detailed soil profile studies are also carried out can this give local information due to taphonomic issues, and species cannot always be identified, with often only a gross category of cereal-type pollen being named (Tipping *et al.* 1994; Havinga 1974; 1984). But the presence of cereal-type pollen, charred cereal grains, and textural pedofeatures in soil thin section profiles has been used in the past to suggest site-specific cultivation (e.g. French 1990, 308). Beyond thin sections, soil pollen has also been investigated regarding ancient environment and land use (Dimbleby 1961; 1985), including slash-and-burn agriculture (Vuorela 1976). In some cases, inclusions of fine charcoal and phytoliths, with burnt soil fragments and soil disturbance indicators, have been related to stubble or grassland burning, or spreading of occupation materials on fields (e.g. Macphail 1990b, 193).

Phytoliths are found in all but the most acidic soils (Smithson 1958; Geiss 1973), and have been studied as site-specific indicators of environmental history and change (e.g. Bobrova & Bobrov 1997; Madella 1997). There has been much work carried out on the identification and taphonomy of ancient cereal phytoliths, especially of rice (e.g. Fujiwara 1993; Houyuan *et al.* 1997; Goldberg *et al.* 1994, 244, 248-9, 255; Courty *et al.* 1989, 115; Macphail 1987b). Phytoliths are regularly found in thin sections of ancient buried soil horizons, where they have been related to different vegetation types, including cultivated plants if they occur with other indicators of these (e.g. Gebhardt 1993, 335; Macphail 1987c). Most identification is based on phytolith assemblages or articulated examples, such as may be found in wood ash deposits and in coprolites (e.g. Murphy 1992). Since phytoliths are said to be physically destroyed by cultivation (Macphail 1987b), they are not expected to be particularly useful indicators for the location of site-specific tillage (but see Alexandrovskiy & Gol'yeva 1997, 269-72).

Weeding of cereal crops is a standard arable activity, although it is not always practised (Russell 1988, 128). Weed control may be greatly influenced by means of tillage. Following Spoor's (1969b) outline regarding weed control through tillage, which distinguishes between shallow-rooted, deep-rooted and rhizatomous weed types, it should be expected that standard ancient ard tillage could be used to control shallow-rooted weak weeds. Shallow-rooted stronger weeds and, possibly, rhizatomous species might be affected by multiple tillage passes; perhaps one of the practical advantages of criss-cross ard tilling would be its effect on shallow-rooted strong and some rhizatomous weeds. When discussed in the literature, however, most weed removal in prehistory is expected to have been done by hand or with the use of hoes. Often such types of weeding do not entail digging to any great depth, meaning that weed roots may not be removed, especially if pulling, hoeing or ard tilling is the method involved. Shallow ard tilling may have little effect on perennial species reproducing vegetatively (Stevens 1996, 191-2; Behre 1981).

Tilling using a mouldboard plough leads to complete inversion of the soil profile. This should have a great impact on weeds, which are both buried and exposed (Spoor 1969b, 17). Perennial weeds, whether deep- or shallow-rooted are greatly influenced by the root exposure created by using turning implements (c.f. Behre 1981). It has been suggested that tilling episodes carried out during fallow periods in extent bush fallow systems are used to keep land free of weeds (Roberts 1996, 127).

Harvesting of cereal crops is proposed to be carried out using either sickles or hand-stripping in prehistory (Rees 1979b; Russell 1988, 40; Stevens 1996). No ancient cereal/grass harvesting and processing methods are expected to usually have a major impact on the soil, with the exception of the pulling up of crops such as flax (Pals & van Dierendock 1988, 242 after Dewilde 1983). Lentils and chickpeas are known to be uprooted ethnographically, with the entire plant being taken for threshing (Harlan 1995, 226); uprooting has been discussed as a grain harvesting method by Rees (1981, 71-2), although this is not supported by certain prehistoric seed evidence (Stevens 1996). The digging required to harvest root crops should have a major impact on the soil, which may be similar to the tillage patterns expected from using implements such as hoes and spades.

Cultivation processes: long-term strategies: fallow and rotation

Pollen remains and some soil studies suggest fallow systems of some sort were widespread in prehistoric Europe (Barker 1985, 52; Rowley-Conwy 1981). There are many types of long-term strategies in cultivation system, and on top of the evolutionary progression between these proposed by, e.g., Boserup (1965), certain types are related to and more suited to certain geographical situations, even before social situations are considered.

For example, in central and northern Europe, cyclical agriculture with a slash-and-burn or swidden system has long been suggested as an ancient long-fallow method (Childe 1929; Clark 1952, 91-107; Bogucki 1988, 81; Tvengsberg 1996; Steensberg 1955; and see Bogucki 1996). In other areas, such as the Wessex region of England, some authors have suggested that empirical evidence for cyclical agriculture does not exist, and that swidden farming is 'unsuitable' and 'unnecessary' (e.g. Rowley-Conwy 1981; Mercer 1981a, xi; Bogucki 1996).

Slash-and-burn cultivation is often thought to leave no 'obvious evidence' *in situ*, and its ancient existence has been interpreted mainly from pollen analysis (e.g. Fowler 1971, 159-60), combined with regional or local charcoal remains (e.g. Huttunen 1980) (although see the discussion above for some localised exceptions). Regional environmental evidence cannot easily be integrated into land-use studies in landscape archaeology (it does not provide evidence tailored for small-scale spatial analyses – Edwards 1982; Thomas 1990, vii; Brown 1997, 133; Butler 1995, 20), but it does provide a context for localised evidence, such as that from buried soils (e.g. French *et al.* 2007; Gebhardt 1990, 76-8), or settlement indicators of agricultural practices (e.g.

Stevens 1996). Even where remains such as charcoal are directly associated with site-based tillage features, there are cautionary tales regarding the relationship of these to slash-and-burn practices (e.g. the charcoal-filled purported ard marks under Sarnowo barrow 8 in Kujavia, which were later suggested to be burnt planks (Whittle 1977, 206; Andrew Sherratt, 1995, pers. comm.)).

Soil micromorphological studies of hoe-dug cornhills in North America have identified fallow episodes (Currie 1994; Mrozowski 1994, 57), and some studies of buried soils in England have suggested rotation or shifting systems (e.g. Limbrey 1992; Macphail 1986a; 1990c; Allen 1992). The identification of rotations and swidden systems archaeologically requires either that they are seen to influence one location over the long term (e.g. arable-grassland rotations on one piece of land), or, through regional data, the 'movement' or alteration of agricultural land uses across the landscape. Such land-use practices have usually been interpreted through other environmental archaeological evidence, and not from the soil itself (e.g. Jones 1988; 1989; Stevens 1996; Smart and Hoffman 1988; Byrne & Horn 1989; Allen 1994; Pearsall 1978; Pearsall & Trimble 1984). However, based mainly on phosphate evidence for poor soil fertility and the later development of peat, Nielsen (1984, 193) suggested both fallowing and manuring occurred on the Iron Age fields at Store Vildmose, Denmark. And Prøsch-Danielsen and Simonsen (1988) suggested shifting pasture and arable land use at an Iron Age site in Norway on the basis of combined pollen and soil chemical analyses.

Summary

Soil micromorphology shows great potential as a technique for describing land-use practices and changes over time. The method has many advantages over more popular approaches (e.g. pollen and molluscan analyses): it is site-specific and has an internal, independent relative chronology; it entails examination and integration of several different aspects/lines of evidence; it employs several scales of analysis and demands an understanding of both environmental and cultural impact; it enables both small- and large-scale environmental reconstruction; it promotes the exploration of the relationship between nature and culture, and of archaeological approaches to this division. However, in many cases there is ambiguity as to the relationship of the soil micromorphological evidence to specific ancient land uses; in particular, micro-indicators related to arable farming may also be related to other disturbance or settlement activities. Interpreting ancient tillage, or any type of land use, in a modern (including buried) soil profile depends on a detailed understanding of the soil type in question, the immediate environment and potential changes in soil formation processes over time, in addition to an understanding of how specific cultural activity directly and indirectly impacts upon these processes. In the above discussion, no distinction was made between different soil and environmental contexts, although features specific to certain soil types were mentioned. The focus has also been on those cultivation indicators that might show up in soil thin sections, due to the focus of the research presented in this volume. In many cases, a cultural activity such as cultivation is marked in a soil through inferred 'anomalies' in the model of how that soil should develop over time, often using comparatives of interpreted uncultivated corollaries, experimental archaeological studies, comparative approaches, as well as theoretical reconstructions of long-term 'natural' pedogenesis and ecosystem change under changing environments (e.g. Courty *et al.* 1989; Goldberg and Macphail 2006; R. Macphail pers. comm. 2011).

The chapters that follow present an experimental approach aimed at better elucidating the processes of how ancient soil tillage features and present-day soil microfeatures are related to each other, and can be used to identify and study ancient arable land use. In this volume, the anomalies studied are those that can be shown to directly relate to tillage implement impact.

Chapter 3 Understanding ancient tilling implements through soil disturbance patterns

Introduction

Ancient tilling is often identified in the soil by the presence of certain disturbance patterns (e.g. ard marks). As the type of soil disturbance pattern created is also one of the main means of assessing implement design in agricultural engineering studies, there is a great deal of information available on soil features created by various types of implements, in many soil types and under many tillage regimes (e.g. Studman & Field 1975, 131; Schafer & Johnson 1982; Spoor 1975). The complex patterns of disturbance are, however, not fully understood either in idealised or non-idealised soils (in the field) (Studman & Field 1975, 131). An additional complication for archaeological application is that the models created for modern implements can only be used as templates for understanding how ancient tools worked.

Archaeological studies have demonstrated through experimental applications that replicas of ancient implements may produce patterns that are predictable from the modern soil-engaging implement work. Studies of soil disturbance patterns created by reconstructed prehistoric ards and historic ploughs have been carried out in the field, for example by Hansen (1969), Lerche (1984; 1994), Reynolds (1967; 1981) and Aberg and Bowen (1960). A previous experimental study of this aspect which included a soil micromorphological component was conducted by Gebhardt (1990), who was able to show a relationship between implement type (*i.e.* specific patterns of soil alteration) and soil structure, as mentioned in the previous chapters.

In general, a soil-engaging part of a tillage implement has two main effects on the soil through which it passes: the share/blade loosens the soil above it and in front of it, and at the same time compacts the soil immediately beneath it and to the sides as it moves. In this description, compaction is defined as the process by which soil is pressed down, 'such that there is a decrease in the volume of voids between soil particles or aggregates' (European Commission 1995-2010), and the term will be used in this fundamental sense throughout this volume, although bulk density analysis was not possible on the microfeatures discussed. Soil loosened by the implement is deposited as ridges, piles and within feature fills. The compacted zone includes implement mark feature cuts and the soil immediately underneath or next to each cut. Implement disturbance patterns are thus marked in the soil by: 1) compaction and friction zones, including the 'gloss' (Lerche 1982, 113) seen at the base or sides of impact zones, and subfeature distortion characteristics created by downwards and sideways pressure, and 2) loosened zones, including fills, 'dragging zones' (Hansen 1969), ridges, etc. (Table 8).

There are many types of digging and tilling implements, all of which produce specific kinds of soil disturbance. In this volume only types which have been related to archaeological contexts in northwest Europe will be discussed. Ethnographic and historic data have frequently been applied to studies of ancient tillage implements, on the assumption that the impact of tillage with any particular implement type on soils is relatively constant, varying mainly with factors such as the draught force needed to till with larger ards or on more difficult soils (Russell 1988, 40). For the purposes of this discussion, implements are divided into type based on patterns of soil disturbance, and not on specific design or material, although it is recognised that both may influence resulting soil disturbance. The shares/blades and sole are the main working parts of an implement in the soil (Šach 1968, 20), and it is these soil-engaging parts that form the focus of the following discussion.

Tillage mark features have occasionally been studied regarding microstructure and organisation. As discussed in the previous chapter, soil horizon and profile traits are rarely associated with tillage in a clear cause-and-effect relationship – there are many other influences to consider. Often it is only or mainly through the existence of a combination of horizon features, such as 'dirty' clay coatings, charcoal, phytoliths, structural indicators etc. (e.g. Davidson *et al.* 1992, 57), that a case for tillage can be made in the absence of other lines of evidence (e.g. ard marks). Tillage mark features themselves have specific organisational characteristics which can be defined micromorphologically, and it would seem as likely that relics of these microscopic traits would survive archaeologically as tilled horizon indicators. Indeed, the various density zones created in and around tillage marks may form the basis for other tillage horizon characteristics (e.g. Jongerius 1983, 122; Pawluk 1980).

Only four implements from ancient agriculture have seen any great focus in archaeological studies: two 'pushing' implements – ards and hoes – and two 'turning' implements – spades and mouldboard ploughs. There are three main defining criteria used by previous researchers to distinguish implement type in archaeological contexts using soil micromorphology: structural characteristics (particularly aggregate shape and size), depth of occurrence of microfeatures related to tillage, and evidence for inversion. All of these criteria see support from macroscopic evidence of soil disturbance patterns associated with the various implements (Table 9).

Table 8 Tillage implement impact zones and disturbance patterns

Pattern	Location	Cause
Mechanical compaction zone (including 'gloss' – Lerche 1982)	Base and sides of impact zone, under impact zone (distortion)	Downwards and sideways pressure from implement moving through soil, compressing aggregates and particles (Lerche 1994, 106); all implements should create compaction zones
'Dragging zone' (Hansen 1969)	Base and sides of impact zone	Deposition of fine particles and aggregates taken up onto a forward-moving implement
Concentration of fine fraction/density-based compaction zone in fill	Base of furrow – in cut, immediately above or below cut	Deposition of fine materials from tillage break-up – trickle-down
Loose fine and coarse angular-subangular blocky and rounded peds in fill	In furrow or mark	Deposition of surface and subsurface aggregates through tillage
Loose fine-medium subangular-angular blocky peds and fine fraction	In ridges or hills	Deposition of surface aggregates from tillage

Ards and ploughs

The focus in this volume is on prehistoric implements, but archaeological ploughsoils (mouldboard), and therefore ploughs, are also briefly described here. Ards are traction tilling tools which push soil up and to the sides without systematically turning it upside down. They are primarily mixing tools. If the ard is tilted when used, soil may tend to move to one side only (Steensberg 1971-2, 19; 1977, 5; Reynolds 1989, 51; Lerche 1994, 11-2).

Prehistoric ards have been classified into many types based on design (Šach 1968; Steensberg 1936; Bratanic 1963; Dosedla 1984). Ards may be simple tools, composed of one naturally bent branch (crook ards), or composite tools consisting of at least one share, which is inserted through the beam (bow/spade ards) (Glob 1951, 109; Šach 1968, 10). The crook ard is thought to be earlier than the bow ard as a type (Sherratt 1981, 267-8).

The main concern here is with the overall soil disturbance patterns created by the various tillage tools and not the design of these. The broad variations between the major categories of tools regarding shape of soil-engaging parts, depth of use and pattern of soil movement are the only details which will be discussed here. In this respect, there are two further categories of ard design which are important: those with and without soles (Šach 1968, 4-5, 11; Dosedla 1984, 44-5), and those with a single straight main share, or with both a straight main share and an arrow-shaped share (Fig. 3). Further variations are associated with share material and dimensions.

Different patterns are expected to be created by straight shares made of stone, wood or metal, based mainly on depth of cut (e.g. straight shares with metal tips should cut deeper). Ard replicas with wooden shares (with and without metal tips), for example, achieve furrow depths of *c.* 12-15cm in sandy soils (Hansen 1969; Aberg & Bowen 1960, 145), which seems to be greater than those produced by stone shares (6-10cm; Rees 1979b, 38). Rees (1979a; 1979b, 13-4) classifies several types of possible prehistoric stone ard shares, based on whether the tools have round or wide cross-sections, and whether they are symmetrical (having a point at each end, potentially reversible) or asymmetrical (having one working point, and a rounded or squared butt end).

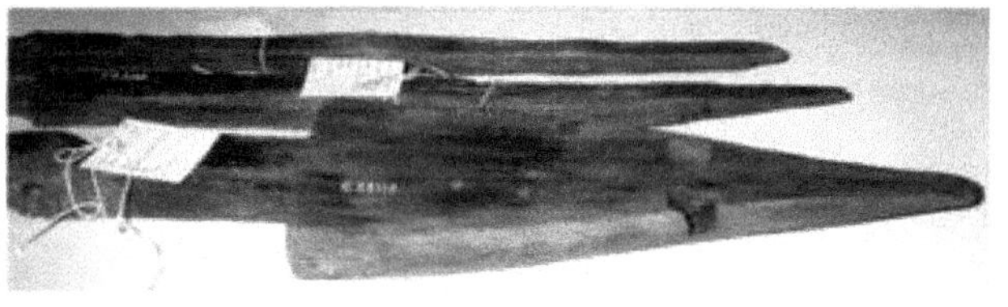

Figure 3 Bow ard replica (Donneruplund) at Lejre Historisk-Arkæologisk Forsøgscenter, above, and arrow-shaped and straight shares of the original Donneruplund ard, below (photographs: H. Lewis, with permission from the Lejre Historisk-Arkæologisk Forsøgscenter and the National Museum of Denmark, Copenhagen)

These differences are expected not to have a great impact on implement depth or angle of working (Rees 1979b, 23), but could have a significant effect on size and shape of soil disturbance pattern. An increase in share length from the early Iron Age is seen when iron shares and sheaths come into use (*ibid.* 33, 53), giving a probable association between iron shares or tips and an increased depth of implement disturbance. The use of an iron tip may be related to the relatively large disturbance zone seen in Aberg and Bowen's experiments (1960) (15cm deep, 25-30cm wide furrows). Experiments with ards show that sometimes relatively deep furrows can be created (e.g. >15cm at Butser Ancient Farm in England). These furrows have been considered to be too deep for immediate sowing, and it is suggested that if archaeological features with such depths are indicators of tillage (as opposed to clearance, for example), some form of field levelling may be indicated (Reynolds 1979, 61).

Experiments with reconstructed ards (e.g. Aberg & Bowen 1960; Hansen 1969), and the results of field studies of archaeological features suggest that ard furrows are normally at least 15-20cm apart from each other and are often as much as 25-30cm apart. It is suggested that they may have been even further apart where irrigation and frequent weeding were necessary (Thomas 1970, 13; Jones 1986, 55; Rees 1979b, 82), and much closer spacing is also known (e.g. 6-26cm at Lodge Farm, England – Everton & Fowler 1978, 181). Archaeologically, ard marks are usually less than 10cm wide (e.g. Thomas 1970, 13), but experimental ard marks may be 25-30cm wide (Aberg & Bowen 1960, 145) or wider, and stepped if the implement has a 'mould-stroker' (*binae aures)* on it (Lerche 1994, 11; Steensberg 1986; Leser 1931; Rees 1979b, 69). The maximum width of an ard mark is equal to the width of the tilling body, which includes the share(s) and any parts which enter the soil (e.g. the sole) (Šach 1968, 8; Payne 1947, 91).

Ard marks found archaeologically should not always be expected to represent the entire zone of disturbance, but only the lowermost parts of this zone. For example, the estimated depth of ard tillage at Wolseley St., London, England, based on surviving ard marks, was only 3cm (Drummond-Murray 1994). This is much shallower than is expected from wooden ards, and suggests some truncation of the tilled part of the profile, possibly in this case by the subsequent flooding which preserved these features (*ibid.*). The loss of the upper part of tillage mark features may be related to disturbance, truncation, or difficulties in recognising the less distinct part of such features, which may be very similar in colour and make-up to the surrounding ancient topsoil. As such, depth of implement mark feature is taken in this discussion only to indicate depth of an archaeological feature in profile, and not necessarily the characteristics of either the tilling implement or tillage practice.

The main factor in depth of furrow is the angle between the beam and the sole (see Hansen 1969, 70). This has been connected to the size of the soil failure zone created by the implement, and is related to traction needs (Tanner 1960). Angle of implement entry into the earth has been discussed regarding the engineering and reconstruction of ards, with most prehistoric shares said to enter the soil at angles of 30-38.5° (Rees 1979b, 33; Hansen 1969). Besides this, depth of impact varies with the length of beam, size and number of draught animals, yoke attachment height, and the operator's pressure on the handle (Šach 1968, 8; Lerche 1994, 12). Soil characteristics (texture, density, stoniness, root matting etc.) also affect ard movement.

Both Reynolds (1981) and Hansen (1969) found that some experimental ard replicas did not create visible ard marks to depths below the topsoil, except where some topsoil was already removed. In Reynolds' case (1981, 99-100), the soil tilled through was only 10cm deep over chalk, and although the ard interacted with the chalk, it did not 'score' it, but only churned it up. Crook ards are suggested to create furrows only 5-10cm deep, which may be too shallow to interact with many sub-topsoil horizons (Steensberg 1973, 115).

Tilting of the ard to one side or the other has been suggested to explain some artefact use-wear patterns. Rees (1979b, 19, 35, 38, 43; 1981), for example, suggests the Donneruplund ard was tilted to the right during operation, but experimental studies with replicas suggest that such tilting may make the ard leave its furrow (Hansen 1969), turn, or even break (Aberg & Bowen 1960, 146; see also Steensberg 1957-8). However, an ard is frequently tilted while one is trying to run it straight. At Lodge Farm, England, it was suggested that tilting could be seen in furrow cross-sections. These were made using an ard without an arrow-shaped share (Everton & Fowler 1978, 180-1), unlike those in the experiments discussed above.

When an ard share enters the soil, disturbance consists of several aspects, varying with implement type, tillage regime and soil type. Experiments at Silsoe Soil Bin with a wooden replica of an Iron Age ard with an arrow-shaped share show that soil is broken up into a mixture of large angular clods and small rounded aggregates using this implement type (Lewis 1998; see below). Gebhardt (1992, 387), however, suggests that the larger clods are mixed with small angular peds. Perhaps this variation in small ped shape has to do with differences in soil types.

The soil immediately above and in front of the shares at all depths is pushed forwards and up the shares, each of which carries soil along on its upper surface, leading to mixing of particles and aggregates from lower and upper layers in the loose material deposited in the resulting furrow and ridge (Lewis 1998). Evidence for this type of implement movement is seen in the wear patterns on stone and wooden prehistoric and Roman period shares, with most wear seen on the top face and on the sides, and only restricted, but distinctive, wear patterns seen on the end of the lower face and the sides of the tip (Rees 1979b, 18-9; Hansen 1969). The greatest impact regarding soil movement is thus caused by the upper face of the implement, which both pushes and lifts. However, soil does flow underneath the share(s) as well, as has been demonstrated with modern implements (Studman & Field 1975).

With the arrow-shaped share, a large zone of disturbance occurs in front of the implement (Lewis 1998), which might be broadly compared with Spoor's (1969a) wide tine soil failure (disturbance) pattern (the width of the cut is much larger than the depth). Narrow tine failure, where a wedge of soil moves forward and up the share without the extra 'surcharging effect' created by wide tines, occurs where depth of implement impact zone is larger than the width (*ibid.*), and might be expected from ards with only a main straight share. In narrow tine failure, the soil 'wedge' carried on the implement originates from the base of the plough zone (*ibid.*), and this may produce horizon mixing.

The fill of an ard furrow is thus a mixture of material from the entire depth to which the share works in the soil, with soil from the lowermost layers moving up into the uppermost by riding up the share(s). This mixture of different depths of soil frequently occurs only in the similar soil materials of the very uppermost horizons, making it difficult to see, especially with the naked eye. If, however, the implement reaches a visibly different horizon or layer, ard furrow fills will include either sub-topsoil material, or topsoil moved down into a lower horizon, and be clearly visible upon excavation, as ard marks. In some cases, the action of the main or straight share may create an internal feature called the 'dragging zone' (Hansen 1969, 80-1). The influence of the 'dragging zone' has been held to hinder identification of implement share type from soil features, although differences can be recognised in soil disturbance patterns created by arrow-shaped and straight shares.

An arrow-shaped share is said to create a wider and often V-shaped furrow, while a straight share creates a narrower and U-shaped furrow, although apparently only one of every 20-50 marks created will demonstrate the shape of the share, and tilt and depth of ploughing can influence this (Hansen 1969, 82; Nielsen 1970; Rees 1979b, 81; Everton & Fowler 1978, 181).

In modern studies of subsoilers, shallow ploughing (up to *c.* 30cm depth maximum) with any shape of share in almost every soil type has been shown to produce 'crescent failure', in which soil is moved forwards, to the side and upwards. It is problematic to distinguish between share shapes when this type of disturbance occurs. Deep ploughing, however, leads to soil flow forwards and sideways only ('lateral failure'), and creates a compaction zone at the base of the cut which reflects the shape of the implement (Spoor & Godwin 1978, 245; Payne & Tanner 1959). Although the implements tested disturb at a much greater depth than ards, the implication that it may not be possible to distinguish between shapes of actual shares used in shallow ploughing is interesting and merits consideration when discussing the individual implements involved in creating features such as ard marks.

Ard furrows can be distinguished from plough furrows in the field by their cuts, fills and the overall organisation of the field (Lerche 1994; Lewis 1998; 2002). Ploughs are asymmetrical traction-tilling implements, which have at least a mouldboard attached (Fig. 4; Lerche 1994, 11; Glob 1951, 109). The cutting action is produced by a horizontally-oriented share in front of the plough body, which undercuts the furrow, after a coulter mounted in the beam has vertically cut the furrow. The mouldboard lifts the resulting slice, and turns it upside-down into the previous furrow (Šach 1968, 5; Lerche 1994, 12; Leser 1931, 159; Rees 1979b, 61).

Mouldboard plough furrows are said to be twice as wide as the share used, and are often 25-30cm wide (Lerche 1994, 271). Proposed Roman plough marks possibly related to road construction in Hampshire and Surrey, England, were 2-20cm deep and 5-20cm wide, with a V-shaped, asymmetrical cut. They were spaced 12-50cm apart (Fasham & Hanworth 1978, 175-6). Pre-Medieval plough marks in Denmark have been described as averaging 15cm wide, 3-5cm deep, and having both U- and V-shaped furrows (Noe 1976, 59). Historic buried plough-tilled horizons around Moscow, Russia, were typically 7-15cm deep, with some topsoil truncation (of up to 10-20cm) (Alexandrovskiy 1996, 330-2).

The fill in a plough furrow is an inverted slice from the neighbouring soil. As such, in situations where several horizons are ploughed, an inversion or partial inversion in horizons should be visible (e.g. Gebhardt 1990; 1995). Fragments of lower horizon material seen in an old topsoil layer may thus represent a specific zone of plough mark fill (Noe 1976, 59). Absence of such inversion features has led to the interpretation of historic period ard use in some cases (e.g. Madsen 1980, 38). The inversion of soil is said to be important in weed control, and for burying crop residues and structurally damaged surface soil (Spoor 1969b).

Other attachments to ards and ploughs include possible precursors to the mouldboard, such as earthboards (to cover broadcast seed, cut ditches, break clod after sowing), and groundboards (to cover seed, create a ridge-and-furrow effect for drainage, cut deeper furrows). The influence of these 'ears' on the soil disturbance pattern includes cutting a deeper furrow and creating a wider disturbance zone (Rees 1979b, 62-5; Dosedla 1984, 45). It seems unlikely that the use of these would be identified in the field, unless examples were found with obvious organisation, such as seen in the schematic cases presented by Dosedla (1984, 45). Where relatively deep furrows or wide furrows are found other possible causes for such shape include those mentioned above, multiple tilling passes in each furrow, or even the use of another implement, such as a traction spade.

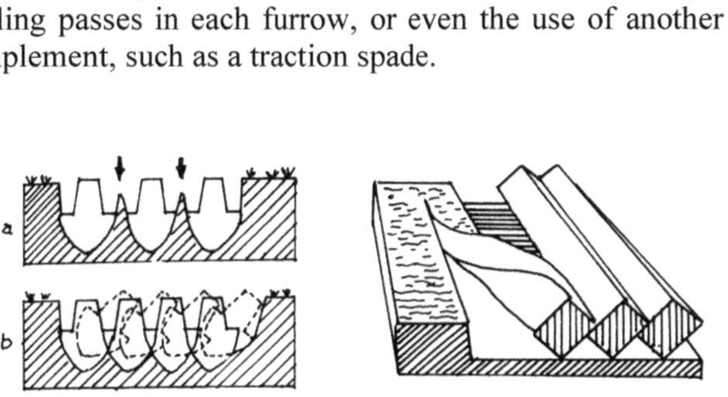

Figure 4 Typical ard (left) mouldboard plough (right) soil disturbance patterns (after Bowen 1963, 8)

Table 9 Summary of predicted field characteristics of furrows and marks created by major tillage implement types (after Lerche 1984, 147-50; 1994, 142, 271; Hansen 1969, 80-1; Rees 1979b, 39, 81-2; Thomas 1970, 13; Jones 1986, 55; Nielsen 1970, 159; Noe 1976; Currie 1994; Steensberg 1957; 1979)

Implement	Dimensions and shape of marks in plan and profile	Spacing between marks	Fill organisation	Cut characteristics
Ard	Main straight share only: <10cm wide, 6-13cm deep (deeper and wider with iron tip), U-shaped; with arrow-shaped share: wider furrow, V-shaped	15-20cm, up to 25-30cm	Fill is mixture of material from all depths of working; subsoil aggregates pulled up into main fill; 'drag zones' may exist, marking the line of passing of the main share	'Gloss' on base of cut, may be sporadic; uneven bases
Mouldboard plough	2-20cm deep, 5-30cm wide; U- and V-shaped; asymmetrical furrows	12-50cm	Inversion; topsoil in sides/base of feature, subsoil at sides/top	'Gloss' on base and side (with the mouldboard) of cut
Spade	Triangular, oblong or rectangular normally; D-shaped, 10-27cm along straight side, 5-20cm deep; V- and U-shaped in section	Variable	Inversion; surficial features at base, subsurface fragments at top	A steeply-sloped side and a gradually-sloped side; rounded or pointed bases
Hoe	North American examples: 7cm wide, 5-6cm deep	Unknown	Mixed fabrics from upper and lower horizons, internal fine fraction translocation	U-shaped

Soil micromorphological indicators which have been specifically associated with ard tillage include the occurrence of medium angular blocky aggregates (Macphail *et al.* 1990a, 61) (especially where these show some vertical or diagonal orientation), mixing of fabrics within the tillage zone, and textural pedofeatures occurring at a (usually inferred) depth of 10-12 or 10-15cm (Romans & Robertson 1983a; 1983b; Gebhardt 1990; Macphail *et al.* 1990a, 61), and weakly-compacted silt panning associated with the base of tilling (at 6cm depth in Macphail *et al.* 1990a, 61). Rounded peds have also been seen (*ibid.*). Compaction has been suggested to not occur with such implements (Carter & Davidson 1998); this has been mentioned above and will be brought up again in later chapters.

Ard marks and overlying 'ploughsoils' have been associated with all of the tillage indicators mentioned in Chapter 2, with the exception of those specifically associated with inversion. Fisher and Macphail (1985, 98), for example, relate a plough pan with charcoal inclusions to 'dusty' clay coatings and to ard marks at West Heslerton, England. French and Marsh (1999; French 2003) describe what may be a possible furrow from Welland Bank Quarry, England, marked by peds oriented at 45° to vertical, composed of lensed very fine sand, dusty clay and fine organic material. They suggest that they may represent 'successive fine infills' of a furrow, although crusting could not be ruled out.

Micromorphological characteristics specific to mouldboard plough tillage as discussed so far are those also seen with ard tillage, but with more organic fills (*versus* the surrounding matrix), and the presence of inversion features reflecting plough disturbance patterns. Indeed, inversion features (including mixing) and depth of textural pedofeatures have been cited as the main predicted indicators (Davidson & Carter 1998, 830). A diagram of a Roman-period plough mark sampled for soil micromorphology was presented by Macphail (1998a, 553; Goldberg and Macphail 2006, 204-7), related to microfeatures composed of phytoliths mixed with fine charcoal and organic matter, as well as to possible manuring evidence (also see Lewis 1998, app. I). The previously-mentioned 10cm-long 'comma shaped' zones of Bt material in the Ap horizon at Malguénac, France, probably represent a specific part of the fill of mouldboard plough mark features (Gebhardt 1990; 1995). In addition, specific textural pedofeatures (e.g. agricutans) and very strongly disrupted and strongly-oriented fabrics have been found in modern plough soils.

Spades

Most ancient spade marks are found in ditches, associated with earthworks, or at the edges of fields, demonstrating the use of this implement for digging, including localised clearance, possibly of turf (Thomas 1970, 14-15; Lerche 1977, 121; Evans 1970, 3). Double paddle spades found in Denmark dating to the Iron Age are thought to be comparable to ethnographically known digging spades, and to have been used for construction (Lerche 1977, 113-4, 119). Examples of digging demonstrably for cultivation purposes are relatively uncommon in the prehistoric archaeological record of northern Europe.

Spade mark features may be very distinctive; although varying in shape and depth with implement shape, dimensions and use, they normally present certain characteristics: a steeply-sloped side marking point of implement entry, with an opposite more gradually-sloping side showing the motion of digging; an overall rounded, oblong, square or triangular shape in plan; and the presence of subsoil fragments in the upper fill (e.g. Lewis 2002). Bronze Age spade marks in England have been described as being D-shaped, 22-27cm along the straight side, and 10-13cm deep with rounded bases, thought to have been produced by a spade with a long handle and a heart-shaped blade (Thomas 1978). Iron Age spade marks at Hengistbury Head, England, were V- and U-shaped in section, 10-15cm long, 5cm deep, and thought to represent a U-shaped blade (Chadburn 1987; Chadburn & Gardiner 1985). As with other marks, it is expected that most archaeological measurements represent the minimum disturbance zone due to later truncation. It should also be noted that the depths seen fall within the same range as those for ard tillage with wooden and stone shares, and that it has sometimes been difficult to distinguish between spade marks and hoe marks in the field – for instance the pre-15th century AD marks found in Red Square, Moscow, which were 10-15cm long lenses or crescents filled with material from an overlying ploughsoil (Alexandrovskiy 1996, 332).

Spade tillage may be envisioned as occurring in several ways. Digging may be used to create a tilth by turning over and breaking up an entire topsoil layer, or by creating ridges or mounds/piles for planting, e.g. in the case of narrow-ridge or lazy bed fields (see below). In the latter case, any spade marks seen archaeologically should be assumed to have been accompanied by ridges or piles located nearby. These piles may be expected to at some point have had internal organisational characteristics representative of the dumping of small volumes (slices) of soil on top of one another, in layers. Prehistoric ridged fields said to be created by spades include Glenree and Belderrig, Ireland (Fowler 1983; Caulfield 1978), and possibly Shaugh Moor, England (Collis 1978, 24). Double-digging, in which soil is turned back into the hole from which it came, resulting in an inversion of the profile within each spade mark, can be carried out to a spatial extent whereby it influences the entire topsoil horizon.

Traction spades (digging motion as a normal spade, but pulled up by a rope) can be used to produce furrows, with soil being piled on one side of the furrow only. These are said to be adapted to paddy farming, but there may be some European examples of use of similar tools (Steensberg 1971-2, 19, 21; Sherratt 1987), including possibly the late Bronze Age V-shaped furrows noted by Harding (1964, 13) at Weston Wood, England, interpreted as spade-dug due to their cut shape (steep and square at one end, shallow at the other, with a perpendicular side always on one side, and between 7-10 and 20cm deep).

Planting ditches and pits have been identified from the Roman period in Britain, and associated with horticulture, although it is uncertain what was grown in them (Jones 1992, 23-4). These were possibly dug with spades or mattocks. The marks mentioned above from Red Square were interpreted as indicating the presence of a garden (Alexandrovskiy 1996, 332).

Spade tillage has been distinguished micro-morphologically by depth of textural pedofeature and fabric disruption occurrence. It has been predicted that spade tillage should lead to inversion and fabric mixing (Davidson & Carter 1998, 830), and that it creates coarse angular aggregates (Macphail *et al.* 1990a, 61). A horizon of spade marks found at Hengistbury Head Site 6, England, thought to date to the early Iron Age (Chadburn & Gardiner 1985; Chadburn 1987) are closely-spaced and appear to mark one temporal horizon of digging or possibly tillage, and their micromorphological characteristics have been discussed by Lewis (2002); these include enhanced organic fills in comparison to the soil outside the features, and planar void spaces, occasionally infilled, at cut edges. Other examples of sites at which spade marks have been examined micromorphologically include Culver Street, Whittington and Wandlebury, all in England. At Culver Street, angular and subrounded fragments of destruction material were found mixed in a spade mark fill (Macphail 1992b; Crummy *et al.* 1992). At Whittington, mixing of burnt soil and brick earth fragments with biological peds and charred plant remains was seen in the fill of a mark (Richard Macphail, 1997, pers. comm.). At Wandlebury, spade marks were delineated by void spaces (cuts) infilled with neoformed sparite crystals, and fills showed mixed fabrics (Lewis 1998; 2003; French *et al.* 2004).

Hoes and mattocks

Hoes are used in many parts of the world for tilling, planting and weeding, as well as for harvesting tubers (Lustig-Arecco 1975, 36). The type of tillage normally carried out with a hoe alone is similar to 'drilling': a depression is scooped out of the soil, with disturbed soil pulled backward (towards the operator), seeds are dropped in and covered up. This type of tilling creates a field of small 'hills'. Hoes are also involved in tillage in the creation of mounds and ridges for planting through scooping soil up from the sides and placing it in the centre of a ridge/mound. This creates layers of topsoil in the ridge and small furrows between ridges/mounds (Harlan 1995, 219; Isaac 1970, 39). Later prehistoric hoe-created ridged fields in North America consist of 10-40cm high and <2m wide ridges (Gallagher 1989, 575-6). Hoes are often assumed to be used in conjunction with another implement (e.g. ard or spade) in Ap horizon creation, and are also associated with weeding operations.

There are many types of hoes, creating soil disturbance patterns which vary in shape and depth. Shallow hoeing would be very difficult to identify from soil disturbance, even in modern soils, but many ethnographic examples suggest hoeing for tilth creation often entails relatively deep disturbance, sometimes as deep as standard wooden ard tillage, and often at least as deep as predicted for ancient stone ard shares (Rees 1979b, 38). In addition, most hoe-based tilling systems produce hills or ridges. As with ard tilling, however, obvious archaeological evidence of hoeing is unlikely to be identified in the field, except where the hoeing action cut down into a distinct lower horizon.

Interpretations of stone 'ard shares' from the north of Britain as mattocks have been refuted on the basis of wear mark patterns by Rees (1979b, 39). Similar types of 'shares' have been interpreted as hoes (Fleming 1995, 33). The possible overlap in depth of disturbance between ards and hoes, and the difficulties in distinguishing morphologically between stone ard shares, hoes and mattocks makes it problematic to have a meaningful discussion on the place of hoes in British prehistory. This is unfortunate, since hoes form a key part of the Boserup model of evolution in agricultural societies so often cited in interpretations of British prehistory.

The only characteristic regularly associated with hoes micromorphologically in archaeology is depth of textural pedofeature occurrence (e.g. Romans & Robertson 1983a; 1983b; Macphail 1987c), presumably related to associated fabric disturbance. Hoe marks are, however, known from North American prehistory, being found in the furrows between cornhills, and these have been examined microscopically, as macroscopically the soil on hoed hills often appears homogenised and features in the hills are very difficult to see (Currie 1994, 68). Currie mentions that the marks may be too faint to see in the field, or be 'masked by faunalturbation features' (*ibid.*). The marks he describes are found in the 'furrows' or hollows between cornhills. He describes a hoe mark fill as being composed of a mixture of fabrics from the surrounding A and B horizons of the cornhill (through which the feature cuts). This suggests that B horizon material was brought up into the main hill, although no description of mounding features or inclusions is presented. Currie (*ibid.* 69) further suggests that features consisting of vertically-oriented shell fragments surrounded by a distinct soil fabric, found in the centre of certain cornhills sampled may represent planting activity, but does not link these to hoe utilisation. The presence of thin layers of horizontally-oriented shell fragments in the horizons outside of the planting feature (*versus* the vertically-oriented ones inside) suggests that an implement cut through and turned the horizons. The horizons outside of the cut of this feature are described as being dense and high in fine fraction content, while the fill of the feature is said to have an increased coarse sand fraction, humic content and lower fine fraction (*ibid.* 66-67), suggesting possible translocation of fine components within the feature.

Interpreted hoeing has also been studied micromorphologically in buried cultivation ridges at North Mains, Scotland (Romans & Robertson 1983a; 1983b), and Chysauster, England (Macphail 1987c). It is possible that Romans (1986, 131) is also referring to hoe or digging stick tillage regarding the initial cultivation period associated with House 2 at Scord of Brouster, Scotland, although he mentions only that seed drills are likely *versus* entire field tillage, based on evidence that the period of cultivation may have been short, and that the original turf was not completely broken up.

Digging and sowing sticks, and other implements

Classically, the digging stick is a round or pointed stick used for digging up a root crop or uprooting wild tubers (Steensberg 1971-2, 20; Lerche 1977, 117; Lustig-Arecco 1975, 11). This category of implement takes many forms, and marks created by these could be interpreted as spade or hoe marks, as stake holes or even as root holes in the field. Archaeologically-found implements are variously identified as tillage or gathering tools, and there may be some confusion with spears and paddles. Sowing sticks are used to create a shallow seed furrow (Reynolds 1979, 63), which presumably would not be identifiable in most cases, or confused with horizontal rooting where found archaeologically.

Rakes and harrows are known from the Romano-British period (Rees 1979b, 484-5), but not from prehistoric times in Britain. Stone 'ard shares' from northern Britain have, however, been alternately interpreted as possibly being fixed into a frame for use in the manner of a harrow (*ibid.* 37). Subsoil marks from rakes or harrows (if any were to go deep enough) might be relatively distinct from ard marks in their spacing. It seems unlikely that marks from rakes or harrows would survive in many circumstances, as these implements may have mainly influenced the upper parts of the topsoil. Nevertheless, Viking period harrow marks have been identified at Lindholm Høje, Denmark (Ramskou 1981, 101-2; Lerche 1994, 148). These had a criss-cross pattern and were proposed to relate to levelling of an associated narrow ridge field system (Ramskou 1981, 101). There is no discussion of how these marks can be distinguished from ard marks – as mentioned above, presumably they were interpreted as harrow marks on the basis of dimensions and spacing, although a particularly narrow spacing is not evident from the plan presented (*ibid.*, 104-5), and Lerche (1982, 114) mentions a variety of spacings ranging from 10 to 60cm.

The preservation and recognition of macroscopic tillage features

The survival of features created by arable practices is influenced in the same way as all archaeological features by factors such as geology, hydrology, soil chemistry, later land use, fill characteristics, and social desires to preserve or alter aspects of the landscape (e.g. Wagstaff 1987, 3; Roberts 1987, 83-85). Soil features directly relatable to prehistoric tillage and digging practices do regularly survive in many environments. Perhaps the most familiar examples from prehistoric contexts are ard and spade marks. Both the profile characteristics and the features associated with tillage are, however, especially vulnerable to reworking, as most are expressed in what was at one time the topsoil, which is has been bioturbated, exposed to erosion, and possibly reworked several times a year or over the course of years through tillage itself (French 1990, 308).

There are two main types of situations where ancient tillage marks and tilled horizons are both preserved and generally identified by archaeologists: as part of 'buried' soils, and when they cut into certain parent materials (chalk or sand especially) in unburied conditions. 'Buried' in this context does not mean a sealed horizon or profile, but that the horizon in question has been covered by a relatively deep overburden, deep enough to prevent later disturbance at the level of the features, and to discourage much (but usually not all) faunal and root activity. Burial does not, however, create a hiatus in soil development. Indeed, continuing alteration of buried soils (in terms of chemistry, organic matter content, biologically-based disturbance and bulk density) has been shown to carry on at least in the short-term following burial under experimental earthworks (e.g. Crowther *et al.* 1996, Macphail *et al.* 1998).

Recognition of tillage features in the field depends mainly on colour contrast (Fowler 1971, 161; Lindquist 1974, 19) (a characteristic based on soil colour and colour of inclusions, texture, moisture and ambient light), except in the cases of major constructions such as ridge-and-furrow fields. Implement marks are commonly found at some boundary (which they may play a role in creating) within the soil profile, either at the top of the horizon underlying the tilled topsoil, or at the top of the tilled horizon itself, as in the case of topsoil burial by blown sand. It is very difficult to find ancient implement marks within a tilled topsoil horizon, although these do exist (e.g. Nielsen (1970; 1986) describes some at Store Vildmose, Denmark). It may also be difficult to recognise such marks in certain subsoil types (e.g. Evans & Serjeantson 1988, 364).

One curious factor regarding the preservation of plough- or ard marks is that often there is one single horizon which clearly stands out from the others. This has been explained as initial clearance activity (e.g. using a 'rip ard' which would create deeper furrows than a normal ard), ritual ploughing events (e.g. where marks underlie barrows), or ground-levelling prior to some other use. The experience of tilling with an ard suggests that it is not an ideal means of ground clearance, even where only pasture exists (Hansen 1969). Evans (1971b, 46) suggests that the survival of the ard marks at South Street, England, shows that, if these marks indeed represent tillage (as opposed to clearance), cultivation activity was short-lived, even though more than one tilling episode is represented.

It is possible that the boundary creation produced by tillage itself may aid in creating the impression of only one tillage episode. And in some cases a combination of cultural interaction with the soil and ongoing soil processes may create the effect of one tillage episode where in fact several are present (e.g. Lewis 2002). This is discussed further below.

Where excellent preservation exists, and tillage features are visible macroscopically, there would appear to be little need to apply methods such as soil micromorphology until detailed field study of the features is carried out (which is rarely – but see, for example, Nielsen 1986; 1993; Everton & Fowler 1978). An outline method for approaching the excavation, recording and interpretation of tillage features and tilled horizons has been developed from work in the Wallonie region of Belgium (Fechner 1996; Fechner *et al.* 1998). The research presented in the following chapters suggests that these implement marks can also provide vital information for the identification of ancient tillage in situations where macroscopic marks are not so well preserved, such as when soils have seen a period of exposure before burial, or have remained unprotected since ancient alteration by tillage.

Dating of tilled soil horizons and tillage marks is problematic for a number of reasons, not least the degree of disturbance which is inherent in their production and in the biologically-active layer in which they are initially located. Nevertheless, in some special circumstances it has proven possible to obtain absolute dates. Carter (1993-4), for example, reports on the radiocarbon dating of narrow-ridged field systems from several sites, based on charcoal concentrations and the basal peat layers overlying some systems. He discusses some of the interpretative problems of relating the tillage features to the dated material (the time lag between abandonment of a field and onset of peat formation, for example).

Archaeological tillage layers and features are usually dated relatively (Catt 1991), although some absolute dating is carried out (e.g. Simpson *et al.* 1998; Nielsen 1984, 62; Madsen 1980, 40; Evans 1971b, 34, 52; Ashbee *et al.* 1979; Wintle & Catt 1985; Courty *et al.* 1996; Scharpenseel & Becker-Heidmann 1992). Most tillage marks are not usually fully excavated, and included artefacts are not generally seen as good dating indicators due to the disturbance inherent in the system. Most tilled horizons and tillage features are thus dated stratigraphically, although often this is only by physical relationship with overlying or underlying features (*terminus post quem/terminus ante quem*) (e.g. Macphail *et al.* 1987).

Field tillage patterns

Fields may be defined as areas of land devoted to agricultural production. Arable land use normally occurs in fields or plots. Arable fields are cultural constructions – they are monumental in scale and form part of the built landscape. But, in their productive function, they also require a preservation of 'natural', albeit altered, soil processes. They are created and maintained both through nature and culture, and this duality is necessary in order for them to fulfil at least one (arguably a major one) of their roles – crop (including fodder) growth. They are often active structures over the long term, and the activities occurring within them change over time.

The ard is an implement which does not generally create significant ridges and furrows such as are seen in ploughed or spade-tilled fields (Lerche 1994, 15). An ard-tilled field is usually tilled from one end to another, going in straight lines back and forth. Later tilling episodes (how much later is unclear) may cross previous episodes at right angles or obliquely, creating the 'criss-cross' pattern associated with ard tilling. While there are times when cross-ploughing does not appear to be a formal tilling pattern (Lamb & Rees 1981, 120), in some cases it seems clear that this practice is part of one tillage episode (Nielsen 1993). Where this is the case, cross-ploughing has been interpreted as a technique required to create a sufficient tilth for planting cereals using a wooden ard (Drummond-Murray 1994). One-way parallel ard tilling, on the other hand, is meant to indicate the use of a heavier implement which could easily break up the soil (Rees 1979b, 80). A mechanically-based explanation for cross-ploughing has been produced through experimental tillage with ards. Aberg and Bowen (1960, 147) suggest that it is difficult to produce close parallel furrows using an ard, as the implement tends to 'fall' into nearby furrows. Presumably such close furrows are desirable for tilth creation, and so, to compensate, cross-ploughing was used in their experiments.

The explanation that the practice of cross-tilling is needed mainly or solely for tilth creation is a little simplistic, although there is no doubt that such a style of tilling has an important role to play in soil disturbance. According to Payne (1947, 83) this pattern was commonly used in historic times in the springtime to help in tilth creation through cutting up previous furrows, and was used in bare-soil fallow systems. While certain types of soils (e.g. clay-rich soils) would probably not be disturbed well through a single ard-tilling episode, there seems to be no soil-based reason why a single tillage event would not create a sufficient tilth in some of the more friable soils (sands, loams) in which cross-tilling remains are frequently found. Other possible reasons for carrying out this type of cultivation could relate to weeding, and/or could reflect tradition or even specific tillage rituals (e.g. Rowley-Conwy 1987). It could also be that this form of field creation could relate to some practical function or specific ritual from its earliest inception, and then have been carried to and continued in other areas as a traditional practice. Or perhaps it reflects a widespread religious tradition (as suggested by Tarlow 1990; 1993/4). Later ard-tilling patterns include following the field shape, and tilling in strips. These are thought to show knowledge of patterns used for mouldboard ploughing (Lerche 1994, 15).

Pre-modern tilling with a mouldboard plough was generally carried out in strips; ploughing would start at the middle of a strip, turning the soil to the right side, and then 'circle' around the strips clockwise, turning each slice into the previous furrow. The furrows at the edges of the strips would be left uncovered ('open furrows'). Lerche (1994) has demonstrated by experiment that once this pattern has been repeated several times a visible ridge is produced along the middle line of the strip, and open furrows occur between the strips. This operation, called 'ploughing together', is the basis for creating ridge-and-furrow field structures with a plough (see Lerche 1994). The ridge-and-furrow develops as the soil accumulates in the centre of the strip, being constantly moved from the edges. The furrows become deeper at the same time, and often deepen so much that they extend below the original surface level, and through the topsoil itself into the underlying horizon. One could level out ridge-and-furrow by interchanging 'ploughing together' with a pattern known as 'splitting', where ploughing starts at the edge of a strip, splitting the ridge and creating two open furrows. The latter is seen as being mainly used since the 19th century AD, when better underground drainage lessened the need for ridges and open furrows (*ibid.* 15-20). At Sarup, Denmark, the fill of 'open furrows' was more clayey (presumably due to crusting, soil creep, etc.) and stony (possibly from stone collecting, water erosion in the furrows, or plough action) than the surrounding soil (*ibid.* 142).

The pre-Medieval origins of ridge-and-furrow are poorly understood: is ridge-and-furrow development simply the result of the use of mouldboard plough technology, or related to changes in crops, land division and tenure strategies? How does it relate to the beginnings of open-field systems? Dating is also a problem. Ridge-and-furrow ploughing is normally dated to Medieval times partly on the basis of correspondence between surviving ridged fields and historically-recorded fields spatially, and because earlier settlement patterns do not appear to correspond to ridge-and-furrow systems (Hall 1981; Fowler 1981a). At Haystack Hill in England, however, the association of this type of field structure with a late Iron Age or Romano-British settlement led the excavator to suggest that either the ridge-and-furrow system also dates to this period, or that the settlement form of 'scooped hut enclosures' has a longer and later history than previously thought (Adams 1996, 6). It is, however, also possible that Medieval farmers respected the earlier settlement – the use of prehistoric field boundaries in Medieval field systems has been reported elsewhere (Denison 1998).

The practice of creating ridge-and-furrow is generally associated with broad-ridged fields (5-20m wide ridges) or grooved rig (as broad-ridged, but without obvious ridges, only widely-spaced furrows) (Parry 1976; Dixon 1994; Carter *et al.* 1997, 447). There has long been discussion over whether certain ridged field forms, particularly narrow-ridged (0.5-1.1m between ridges), were created by ploughs or using spades/shovels (e.g. Ramskou 1981, 101; Lerche 1982; 1994, 148; Carter 1993-4, 90; Halliday *et al.* 1981, 56). Spades are known to have been used in ridge creation at least in historic times (Bowen 1963). Cord rig (<1.25m wide ridges) has been ascribed to a combination of ard and spade tillage, and the use of some ridge-and-furrow systems for the production of grazing and not cereal crops has been proposed (Carter *et al.* 1997, 447). This is supported by historical evidence which documents the use of ridge-and-furrow for grass growth and as permanent pasture during post-Medieval times. Such land use may mark a change in practice at the end of the Medieval period (Hall 1981, 26). Case studies of the histories of particular ridge-and-furrow fields and field systems, some of which may date from at least 5,000 years ago, include those by Hall (1981), Caulfield (1978) and Chadburn (1987).

While ridge-and-furrow systems have seen typological classification (Bowen 1963; Parry 1976), the different morphologies presented by spade-tilled and ploughed narrow ridge systems have not been discussed to such a great extent. Spade tillage might produce substantially different ridge organisation when compared to ploughing, although both overturn a slice. Traction spades, which also turn to one side, should still create piled soil, and do not necessarily lead to overturned slices (Steensberg 1971-2, 19, 21).

Hilled fields may be created by hoeing, with piles of soil for planting being separated by furrows from which the soil has been hoed up onto the hills. Hoe marks may be seen in the furrows. Ancient examples of such fields have been studied micromorphologically in North America (see above). Various macroscopic characteristics of hilled fields include a very thick A horizon, created by mounding of topsoil and B horizon material from the furrows between the hills, along with the addition of organic matter (as in manuring or plaggen-type deposits). In addition to the normal secondary processes associated with bare earth and disaggregation, hills in ancient cornfields saw strong colluvial movement (with deposition at the bases of and between hills), crusting and bioturbation (Currie 1994). Presumably spade tillage could also be expressed through hill creation.

Both ard- and plough-tilled fields may possess headlands. A headland marks the point at which the team turns, usually with a lifted implement. Headlands are created through a combination of soil accumulation through trampling, manure addition, rubbish accumulation, and from soil dropping off when the plough is lifted (Hall 1981, 23).

Clearance cairns have been associated with traction-based arable agriculture on the assumption that there would be no reason to clear larger stones from fields for stock rearing or spade tillage (Collis 1978, 24), while a field must be cleared when using a plough or ard. In some very stony cases, however, it has been suggested that stone clearance would have to precede use of fields for pasture as well as cultivation (e.g. Whittle *et al.* 1986, 140), although some authors suggest otherwise (e.g. Provan 1973, 41). Mercer (1981a, xi-xii) suggests hoe or spade tilling at Carn Brea, England, because small clearance cairns are irregularly dispersed (see also Rees 1979b, 39 regarding some Scottish sites). Lerche (1994, 137) associates clearance cairns with ard tilling, as opposed to plough tilling, using historic, archaeological and ethnographic evidence suggesting that stone clearing when using a ridge-and-furrow system tends to lead to accumulations of stones in 'open furrows'. When ard tilling, these drainage furrows do not exist, and stones are generally moved to one part of the field or off-field (Hatt 1949, 44-5; Lerche 1994, 137). There are some archaeological examples of clearance cairns associated with narrow-ridge systems (e.g. Carter 1993-4, 85, 89). These are assumed to have been plough-tilled (*ibid.* 90). In general it seems that the presence of clearance cairns may not be indicative of ard tilling, just of the lack of a larger ridge-and-furrow system, which is often associated with stone rows. Clearance cairns associated with plough marks have also been found in Northumberland, England (Feachem 1973, 339; Jobey 1968). The association of field clearance with burial cairns (e.g. Spratt 1978) has sometimes also been worked into discussions of a perceived relationship between burial and fertility associated with arable agriculture (e.g. Gaffney *et al.* 1995).

Lynchets are linear features created through downslope movement of soil loosened by tillage or clearance on a slope. When a field on a slope is tilled, disturbed soil from the top tends to wash downslope and collect against the lower unploughed edge of the field. This results in a characteristic morphology comprised of a negative lynchet (shallow depression) at the top of the field and a positive lynchet (bank) downslope (Childe 1949, 190; Wood & Whittington 1960, 336-7). Lynchets often develop against walls or other slope-breaking features, some of which appear to have been constructed for the purpose of arresting soil downslope movement (e.g. Whittle *et al.* 1986, 49-50).

Several types of ancient lynchets and terraces are listed in the literature (e.g. see Wood & Whittington 1958-9; 1960; Jackson 1957). Some excavated examples from later periods appear to represent deliberately-created field structures, similar to terraces, perhaps for levelling, creating a deeper topsoil, or to offset future erosion (Wood & Whittington 1958-9; 1960; Wood 1957-8, 22; Lerche 1994, 304). Pre-lynchet surfaces vary, but many of those excavated have been either deturfed or tilled (Fowler 1967, 16-26; Bell 1977; Drewett 1978; Wood & Whittington 1958-9; Wood 1957-8, 22). Some lynchetted fields have ard marks, and this, along with the size and shape of the fields demarcated by lynchet formations are said to indicate traction tillage *versus* hoe or spade tillage (Barrett 1994, 149). This is based on the idea that tilling with ploughs equates to rectangular strips, with ards to square plots (as with many lynchetted fields), and with digging sticks, hoes or spades to rounded or irregular plots (Lerche & Steensberg 1980; Fenton 1994). The basis of these proposed relationships is a combination of ethnographic examples (e.g. Ruthenberg 1980), the amount of work a plough team can do in a day, the proposed necessity for cross-ploughing with ards to create a sufficient seedbed, and the number of oxen used. These relationships have all been disputed (Curwen & Hatt 1953, 61-3; Russell 1988, 40; Payne 1947; Fowler 1967, 25-6). Moffat (1988) suggests that increasing field sizes in the Iron Age and later in England should have a relationship with increased levels of off-field erosion (after Fasham 1980; Evans 1980). However, one proposed explanation for the beginning of lynchetted fields is that they were constructed to offset erosion (e.g. Bradley 1978b; Bell 1981b); in such situations it may be field abandonment (lack of such management, and decreasing intensity of cultivation) which leads to erosion episodes.

Studies of sections through lynchet banks show variable internal structure, with some being homogeneous, while others have faint or quite distinct layering. The layers may relate to original formation, or to post-depositional pedogenetic processes occurring within positive lynchets. Lynchets often cover topsoil layers (Wood & Whittington 1958-9, 164), some of which appear to be turf horizons or 'culture' layers. In other cases they are underlain by ard marks (Fowler 1983, 145), or buried soils with arable indicators (Keith-Lucas 1986, 107), sometimes with compaction noted (Wood & Whittington 1958-9, 167-8). In general, ancient field systems on sloping ground have been assumed to be arable where evidence of downslope soil movement is seen (soil loss upslope, accumulations such as lynchets), and to not have seen arable activity where this is not the case (Courty *et al.* 1989, 130). Romans (1986, 130) suggests that it is possible to interpret ancient depth of midfield ard tillage through estimating 'ratio of midfield plough depth to downslope accumulation' (being expressed through the depth of lynchets/colluvial material gathered against walls etc.), and on this basis he suggests a 9cm depth of midfield ard tillage in field 2 at Scord of Brouster, Scotland. The impact of aeolian erosion or other soil loss is not discussed, and he assumes that the ratio seen from a relatively recent croft field with a 7° slope can be used as a comparison.

Terraces are often associated with lynchets, and have been related to prehistoric arable activity in Europe (e.g. Nielsen 1993; Andersen 1989). They build up through the addition and stabilisation of soil against walls, banks or lynchets on a slope. Terracing is said to lead to a more open soil structure with a greater content of coarse materials and increased infillings with fine materials in lower (sub)horizons. In the Mediterranean, terracing has been linked with an increased rate of leaching (de Olmedo Pujol 1983). Other studies show that terraced field systems produce and retain deep soil profiles, offset rather than prevent erosion, and are often constructed on naturally 'poor' soils which require frequent maintenance to prevent degradation (e.g. Goodman-Elgar 2008).

Field systems are difficult to study for land-use history; for instance it is unclear whether prehistoric field systems in England mainly represent stock control (e.g. Fowler 1981a, 45-6; 1981b; 2000), and/or arable land use (Fleming 1978; 1989b, 64, 66). As many of these fields have seen continued use, only a few vestigial soil profiles exist which could contain information on ancient land use, although where deposits survive in field boundaries such as ditches, banks or lynchets, it is possible to gather some information on environment and land use (e.g. Entwistle 1994; French *et al.* 2007).

'Permanent' fields are associated with certain types of constructions: boundaries such as lynchets, ditches and walls in particular. Where such archaeologically-visible structures are not found, the possible presence of arable or pasture fields is often ignored. Field boundaries such as fences, clearings bounded by woods, hedges, or simple crop change, are generally not discussed, probably because they are archaeologically difficult to locate (but see Bowen *et al.* 1978; Pryor 1976; Holden & Bradley 1975; Ashbee *et al.* 1979, 258-9; 1986, 41, 45; Nielsen 1984, 139, 161; 1993; Allen & Robinson 1993). Moffat (1988) suggests that boundary destruction is not a major factor in surviving ancient field distribution patterns in southern England. Recent evidence of extensive loss of soil depth to recent ploughing, at least on the chalklands, suggests that this assertion needs serious re-assessment on a landscape scale in that region (French *et al.* 2007).

Summary

Experimental approaches have long been popular in archaeology, particularly the reconstruction of 'primitive' implements and experimentation with tool replicas. The focus has usually been on identifying use-wear patterns, describing the impact of tools upon deposits similar to those seen in the archaeological record, and understanding how tillage technology influences and interacts with other aspects of prehistoric culture (Coles 1973; Reynolds 1977; 1981; Lerche 1994; Anderson 1992). In archaeology, and to some extent in agronomy, the study of ancient implements has entailed the use of ethnographic and historical analogy regarding tool use and crop production, and the place of tillage and arable farming in modern cultures and in human history (e.g. Butler 1992).

In order to relate ancient arable land use to landscape, it is important to investigate site-specific land-use practices by exploring the landscape for fields, and then examine fields themselves for indicators of land use and other cultural practices. The study of fields, and especially field systems, has a long history in the archaeology of northwest Europe (e.g. Curwen & Curwen 1923; Hatt 1949; Nielsen 1970, 1984; Feachem 1973; Hall 1981; Peterson 1990; Taylor 1975; 1981; Bradley 1978a) and several studies have attempted to address how land use is integrated with the development of field systems (e.g. Fleming 1978; 1985; 1988; 1989a; 1989b; Lewis 2008; Fowler 2000; Chadwick 1997; Johnston 2005; Wickstead 2007). Where field features such as ard marks are found underneath monuments such as round barrows, it would be interesting to explore the relationship(s), if any, between ideas of changing land uses and landscapes represented by the two types of remains, as opposed to simply appropriating the arable land-use evidence into the monuments (see, e.g. Nielsen 1993, ch. 10; Becker 1992-3, 166-7).

The field as a feature deserves much more attention in landscape archaeology, not least because it forces one to address the perceived culture:nature divide (McGlade 1995), by presenting a spatial and temporal physical entity in which these two categories are melded. Despite this, fields are often relegated to the day-to-day, mundane, 'purely economic' aspect of life (if such a thing exists today or ever existed in the past). The standard archaeological perception is echoed by Bradley (1992, 17) in his contrast between fields and monuments: 'Unlike field systems, monuments have very little to do with adaptation, so they need not take the same forms just because they occupy similar environments. They may not have had fixed meanings in prehistoric society that could be transferred from one region to another...Ceremonial monuments could convey *meanings* that might be subject to local interpretation'. Most prehistoric fields known have not been excavated, and it is thus impossible to suggest their histories in terms of land use, form or meaning, although field remains play a large role in perceptions of ancient landscape (see Lewis 2008; 2010).

Chapter 4 Characterising experimental implement marks micromorphologically

Introduction

One principle behind conducting experiments with reconstructed ancient tillage implements for soil micromorphological study is that it is both possible and necessary to isolate the influences of tillage with such tools on the soil. Previous research demonstrates that this is possible in regard to certain structural characteristics and the creation of a few types of pedofeatures (e.g. Gebhardt 1990). The intention in isolating one or two components is to simplify the soil system to a point where influences can be readily recognised and understood.

In archaeological studies of tillage the isolation of certain features of the soil and/or tillage system is especially important, as the main concern is not with how a soil performs or reacts as a body today under cultivation, but with surviving fragments representing its arable history. Only small relics of the ancient system are normally encountered. It is vital to understand what these fossil structures and features represent in terms of past cultural activity and environment.

Tillage comprises a set of processes which is, however, often quite difficult to isolate in non-modern soils (e.g., Carter & Davidson 1998; Macphail 1998a). The characteristic features of these processes may sometimes be created by other means, and they are relatively easily altered by subsequent activity, especially because the main impact of tillage is in the physically, biologically and culturally most active horizon – the topsoil. It has been suggested that experiments using reconstructed tillage technology for the purposes of soil micromorphological study of ancient arable farming should be especially concerned with creating an accurately replicated tillage method, using soils resembling those worked in ancient times (Carter & Davidson 1998, 541; Davidson & Carter 1998, 827), although the need for and possibility of creating such conditions has been questioned (e.g. Macphail *et al.* 1990a). The location of soils *possibly* similar to those which *may* have seen arable land use in prehistory is, however, only one part of the problem in meeting those conditions. Access to and availability of an appropriately controlled environment in which to conduct experimental studies is also an issue. A situation that would meet both of these conditions could not be found for even one type of soil involved in this project, and Gebhardt (1990) met with a similar circumstance for much of her experimental research.

The replication of ancient tillage systems, while undoubtedly preferable, is an ideal which it has not been possible to achieve in these or any other previous experiments in 'ancient' tillage. Too little is known of each aspect of these systems in prehistory, and even less is understood of how implements, soils, crop types, amendment and other components were combined. However, studies of historic and recent traditional systems (e.g. Gebhardt 1993; Davidson & Carter 1998) can provide an informative backdrop for investigating interpretative and methodological issues regarding ancient arable systems. The following chapter presents the results of studies at three Butser Ancient Farm sites, Lejre Historisk-Arkæologisk Forsøgscenter and the Silsoe soil bin, and discusses where these data fit in the body of previous experimental research. The results of the tillage experiments are described in two parts: feature characteristics (this chapter), and profile and horizon characteristics (Chapter 5). All of the replica experiments were designed and conducted by other researchers, with the exception of those at Silsoe.

Ard tilling experiments

An experimental study was conducted to further characterise the impact of ancient tilling implements on the soil through ard tilling experiments at the Silsoe College soil bin (Cranfield University, Silsoe, UK), and Lejre Historisk-Arkæologisk Forsøgscenter (Lejre, Denmark).

At Silsoe experiments were designed with the purpose of observing the basic soil disturbance patterns produced by a reconstructed wooden Donneruplund ard with wooden arrow-shaped and straight shares (Fig. 3; after Glob 1945), and comparing observed macroscopic soil organisation to features seen in thin section. Use of the soil bin ensured no influence of weather, season or biological activity. Disturbance could be examined with relatively constant original soil structure and texture (fine sandy loam); the former was apedal macroscopically with rare blocky peds at the surface, and microscopically single grain (some capped), intergrain microaggregate and 'rolled' aggregate (rounded aggregates *cum* nodules composed of finer grained components around a coarse sand grain core). Resistance was generally 1500-2500kPa in the area tilled (increasing with depth), moisture content was 7.6-8.7%, and consistence was uniformly friable or soft. The soil bin was prepared to the same standards before each tilling event, in 50mm thick layers. Each layer was compacted by rolling six times using a 700kg roller (after Desbiolles 1993). Lines of powder paint were added at 2m intervals along the bin and at 4-5cm intervals down the soil profile to assess the impact of the ard at various depths, and the lateral distance to which soil aggregates from painted depths were moved. The ard was run using a variety of share lengths, traction heights and angles in the main soil bin at Silsoe. Most of the runs had a yoke height approximating that of a pair of small oxen with a neck yoke (c. 120cm). Draught force was around 60kg, approximating that of two small oxen[1] (G. Spoor, 1994, pers. comm.). All runs were in the same direction, and there were no attempts at cross-ploughing, although repeated ploughing following the same furrows was carried out in some runs. Details on measurements and additional characteristics of the soil used are given in Lewis (1998).

It was predicted that features which might survive subsequent disturbance would be those created by the movement of soil from lower levels up the shares into the surface layer (to be demonstrated by the use of paints at various levels in the soil bin), and the compaction layer underneath the share and foot of the implement. Block samples for thin sectioning were taken from excavated profiles immediately before tilling and after each tillage event, general field notes were made during and after each run, and video footage of the experiments was produced for later study (Lewis 1998). Twenty block samples were taken, including tilled and untilled examples (Table 10).

Macroscopic observations of Donneruplund ard disturbance at Silsoe

The ard produced a wide V-shaped furrow when the arrow-shaped share was used, or a rounded U-shaped furrow with only the main straight share. Furrows ranged from 6-7 to 12-15cm deep and 21-22cm wide ('true' depth and width, *i.e.* the cut of the mark). The latter depths are the same as those seen by Hansen (1969) using a Hendriksmose ard replica (Døstrup type) in sand, and the variation in depth might be attributed here to lower resistance in some parts of the soil bin as compared to other parts (where furrows were shallower). Aberg and Bowen (1960) found that a horse-drawn Donneruplund ard with both shares produced V-shaped furrows 25-30cm wide and 12-15cm deep using an iron tip on the straight share.

The stepped profile typical of an implement with a second, arrow-shaped share was produced, and the types of soil failure patterns expected from both narrow and wide tines (Spoor 1969a) were observed. Wide tine failure is produced as the soil in front of the ard is moved forwards and upwards along a 'failure surface'. Narrow tines tend to create a soil 'wedge' which comes from the soil material cut by the base of the narrow share (the straight share here). This wedge is brought up to the surface along the upper face of the tine (*ibid.* 23, 26).

The arrow-shaped share seemed to act mainly on the upper 2cm of the profile, creating a 'ledge' in the mark's profile; more width/volume of soil displacement occurred when both shares were used simultaneously than when the straight share was used alone. The following effects were noted of using the two shares together (*versus* only the straight share): 1) different profile shape (with a ledge) and soil disturbance across a wider area; 2) a greater volume of soil disturbed, but without much influence on the depth of the furrow; 3) some additional compaction on the base of the furrow, and the presence of a zone of 'rolled' aggregates at the base or sides of the furrow (see below); and 4) with the arrow-shaped share, soil appeared to break up so that there were more small and slightly less angular clods. The distinct outline in profile created by the arrow-shaped share was clearly defined in the soil bin by compaction, rolling and deposition of fine materials at each feature cut (Lewis 1998). All of these macroscopically-visible features could in theory survive in archaeological sites where marks survive to a depth of greater than 5cm.

Table 10 List of soil micromorphology samples from the Silsoe soil bin (see Lewis 1998 for full details)

Sample	Run number, no. of times each furrow tilled	Description
SSB1-2	n/a, untilled	Background, no powder paint
SSB3	n/a, untilled	Background, powder paint
SSB4	I, x1	Side of furrow, 'true' profile
SSB5	III, x1	Base of and underneath furrow, with loose soil
SSB6/1-2	n/a, untilled	Background, powder paint
SSB7	7III, x10	Ridge with loose soil, 15cm ahead of powder paint line
SSB8-9	7IV, x1	As SSB7, on powder paint line
SSB10-11	7VI & 7I, x1	Furrow bases with loose soil, on powder paint line
SSB12	7II, x4	Furrow base 'true' profile, on powder paint line
SSB13	7III, x10	As SSB12
SSB14	7V, x5	As SSB12
SSB15	7II, x4	As SSB12
SSB16	7III, x10	As SSB12
SSB17	7V, x5	As SSB12
SSB18	7II, x4	As SSB12, 1m ahead of powder paint line
SSB19	7III, x10	As SSB18
SSB20	7V, x5	As SSB18

Two basic types of aggregates were produced by these experiments. Angular blocky clods of a relatively large size, averaging 7cm, were found mostly in the ridges, but some were also seen in the ard mark fills. Many of these were cut by the ard at the surface. These aggregates had no precursor in the original structure of the soil. It is uncertain to what extent clod shape here is a function mainly of the soil preparation procedure in the bin or of implement action, although descriptions of similarly large angular and subangular clods were also associated with experimental ard tillage by Gebhardt (1990, 44-6, 63-6). In addition, there were small (<2.5cm), more rounded

[1] The size of the oxen was based on *c.* 4ft high Dexter cattle, which were estimated to weigh *c.* 700-800lbs each (P. Reynolds, 1994, pers. comm.).

aggregates found in the ridges, and frequently also in the loose of the furrow. These originated from various depths, particularly from the surface and from *c.* 5cm depth (Lewis 1998). These types and arrangements of aggregates sound very similar to those resulting from modern tillage using vertical and forward raked narrow tines, which bring larger clods to the surface, and leave smaller ones below (Spoor 1969b, 18).

Using the different colours of powder paint laid at 5cm depth intervals, it was noted that aggregates from the lower part of the implement impact zone were pulled up into the top of the furrow, and that fragments from upper levels dropped down into the furrow fill (Lewis 1998). Mixing of aggregates from upper and lower levels occurred in the loose material both in the furrows and the ridges. Large surface clods tended to be deposited on the outside of ridges, with subsurface material tending to be located on the furrow-side of ridges and in furrow fills. Soil from greater depths was acted upon by the main straight share, riding up the share and mixing with material from upper layers both on the implement itself and upon deposition. Furrow fills contained angular and rounded aggregates in their upper deposits, and a lowermost thin layer dominated by fine aggregates and disaggregated apedal material, with only occasional larger angular and subangular clods (*ibid.*).

The movement of powder paint lines in the soil bin also showed lateral displacement (forwards and sideways) of aggregates from all depths encountered by the shares. Soil moved ahead of the implement from both surface and lower layers following the notion of soil failure zone or disturbance crescent (Payne & Tanner 1959, 312, 316-8), and also upwards and sideways from the point of origin. The distance of aggregate movement should be a function of soil type and of implement dynamics (see Studman & Field 1975 for a discussion of such movement regarding stones). Here, surface aggregates moved on average *c.* 30cm forward, while those at 5cm depth moved *c.* 13cm. The maximum upward displacement observed was 6-7cm for all layers affected. Minimum sideways movement was 2-3cm. Most aggregates from 5cm depth were deposited in the loose soil on the inner side of the ridge. Occasionally, small clods (1-2cm) from this depth were found in the loose in the furrow, and rarely in the loose on the outer side of the ridge. Only rarely was inversion of the aggregates from this depth observed. Overall, from all depths, smaller clods were seen to be moved the greatest distance forward (up to 1m), while movement of larger clods was restricted (up to 30cm). Both showed similar sideways movement into ridges, although larger clods tended to be deposited on the outer side and at the surfaces (Lewis 1998). Whether any of these arrangements would survive in archaeological situations would depend on burial and preservation environments; recognising such movements would depend on there being clearly distinguishable horizons within the furrow depth in question.

The Lejre experiments

It was also possible to sample from oxen training runs for an experimental tillage project with a wooden Donneruplund ard replica at Lejre. The Lejre Historisk-Archæologisk Forsøgscenter was established in the 1950s, and experiments using reconstructed ancient tillage implements have been conducted there for many years (Hansen 1988). It is located in a glacial landscape, with low rolling hills, small lakes and marshy areas. The underlying geology is mixed moraines and outwash sands overlying Zelandian and Danian deposits, and the dominant soil types are brown earths. In the immediate vicinity land-use types are mixed, with areas of meadow, forest and cultivated fields (Marianne Rasmussen, 1995, pers. comm.; Lokal-plan 4-5; maps & sections nos. 1, 3, 4, 8 & 11 produced by TEK-SAM, RUC, Feb. 1989).

The soil in the field studied was an argillic brown earth. The strip used for training had been broken originally using a tractor-pulled harrow (M. Rasmussen, 1994, pers. comm.). Two Dexter oxen (1.2m tall) pulled the ard with a yoke height of 1m (but with a long rope, so yoke height varied). Both shares were used simultaneously, with the main straight share 10-15cm out. The Lejre profiles are comparable in regime to some created at Silsoe, namely those which saw repeated runs through the same furrows (possibly 10-20 or more runs at Lejre – N. Kristiansen, 1994, pers. comm.), and similar characteristics to those at Silsoe were seen in the 'true' profiles observed. However, the Lejre examples were located outdoors and were tilled over a period of several weeks, in a real soil. The repetitive tilling resulted in relatively little soil remaining in the Lejre furrows. Furrows were 4.25cm deep and 12.9cm wide at the top on average ('true' values), making them much shallower than those created at Silsoe, or than is generally expected. The ground was hard, making penetration difficult, and this appears to account for furrow dimensions; earlier tilling training in the Spring using the same replica ard was reported to have created good furrows (N. Kristiansen, 1994, pers. comm.). Two block samples were taken of 'true' furrow bases and the immediately underlying soil. Samples were also taken from an additional plot which had been broken with a modern plough in 1993, ard tilled twice and then left fallow. One sample from this plot produced a further ard mark feature identified in thin section only (see below).

Soil micromorphology study of experimental ard marks

During both of the tilling experiments, samples for soil micromorphology were taken to examine five main issues: 1) the structure of the 'topsoil' resulting from ard disturbance, 2) the movement of different parts of the soil up or down the profile/implement, 3) the organisation of the ridges, 4) the organisation of the furrows and 5) any effects seen in the underlying or neighbouring soil.

Untilled samples SSB1-3, SSB6 from Silsoe

Pre-tilling background samples from Silsoe showed the overall profile to consist of one main fabric, which is a sandy loam. 'Horizons' of 4-6cm thickness were created by the soil bin processor. These had distinct boundaries of fairly straight horizontally-running linear voids, often continuous, but sometimes interrupted by fine aggregates, in which case they comprised lines of interconnecting short planar voids or vugh-like pores. The linear void spaces will be called 'planar voids' throughout this discussion, although they have characteristics of both planar voids and channels (after Bullock *et al.* 1985, 44-5; Fitzpatrick 1993, 115-6). Like channels they were of a roughly equal width throughout their length, and compaction was seen in adjacent soil material, with infilling by fine aggregates common. But they showed strong accommodation of their walls, and were akin to planes in their origin – they were created through the physical deposition of soil and stresses involved in rolling out layers. These horizon-marking voids were mainly 1,500-2,000μm wide, but narrowed to 50μm wide occasionally between deeper horizons. Where powder paint was added to the soil, this normally almost completely infilled the horizontal planar voids (leaving only a very fine central planar void), and extended into immediately overlying and underlying voids (Lewis 1998).

Overall, the macroscopic structure of the untilled soil was apedal, but with subangular blocky (0.5-1.5cm) and elongated subangular blocky or short 'prismatic' peds (2x1cm) in some upper horizons, and a few rounded aggregates (up to 2cm sized) seen throughout. There were several structural components evident at the microscopic scale:

a. single grain – individual sand grains of all size grades, sometimes with cappings, coatings and pendants of fine fraction, and a 'rolled' appearance, round-subangular in shape,
b. intergrain microaggregate – irregular fine crumb-like and very fine subangular blocky microaggregates,
c. 'rolled' aggregates – rounded aggregates composed of finer components of types a-b; some were large enough to be visible macroscopically,
d. subangular blocky and rare granular-like peds composed of types a-c (macroscopically visible), and
e. elongated subangular blocky to short prismatic peds composed of types a-d (macroscopically visible).

Clay coatings, and very fine sand and silt coatings were seen on 60-80% of grains, oriented concentrically, with a low porosity, giving the impression of having been rolled. The large 'rolled' aggregates had a similar orientation of finer components. These various structural features occurred in a poorly-sorted manner, except for the macroscopically visible peds, which were relatively regularly spaced, and only rarely found outside of the uppermost horizon or top of the second horizon. Porosity varied with structural component, decreasing slightly with depth. The impact of the rolling and compacting actions of the soil bin processor on the soil was apparent in several aspects: an overall decrease in porosity with depth (seen in horizon porosity estimates), the increased tendency to infilling and narrowing of the horizon boundary voids, and the increasingly apedal nature of the soil with depth (Lewis 1998).

Lines (<50μm wide) of fine components, appearing as infillings, but deposited in the absence of any leaching, were seen running horizontally and echoing the horizon boundaries marked with powder paint; these linear infillings appear to represent the boundaries of some of the other 5cm-thick soil layers laid by the machine, which were then subsequently compacted. Other characteristics of the background soil samples are described in Lewis (1998) and summarised in Table 11.

Furrow fills and cuts from Silsoe (SSB4-5, SSB10-20)

Four samples were taken with the main furrow fill intact. All show two to three furrow fill layers, distinguished by texture, structure, organisation and porosity (Table 11). In all cases, the uppermost furrow fill (<1cm thick) was predominantly single grain with intergrain microaggregate structure, with a porosity of 30-50%, comprised almost wholly of packing pores. Small (<0.5cm) subangular blocky and 'rolled' aggregates occurred within the fill. These had a low porosity (10-20%), and were similar to the aggregates described in the background samples. The very uppermost part of this fill was too loose to be sampled for soil micromorphology.

Underlying this was the main furrow fill (<6cm thick), comprising large zones of loose fine aggregates and grains as described above, but with occasional 2-4cm angular-subangular blocky clods composed of densely-packed microaggregates (Fig. 5). These clods had no precursor in the original soil structure, and were created by the action of tilling. Other components seen include rare small 'rolled' aggregates as seen in the uppermost fill. Porosity in this main fill is lower overall than in the overlying, due mainly to the presence of the denser, larger aggregates (Lewis 1998).

Where seen, the basal fill was a fine layer (<1cm thick) of medium, fine and very fine sand and silt grains and fine aggregates (Fig. 5). The creation of such 'lenses' of fine materials in a soil bin and in a soil with this texture simply by tilling suggests that movement of disaggregated finer particles down to line the feature cut is a primary process in ard tillage, and does not occur only under the influence of water (leaching); this is discussed further below.

The furrow fills thus showed a combination of new large aggregate creation, the disruption or removal of already-

SSB 4 (1:1 photograph in light box; thin section is 6 x 9 cm)

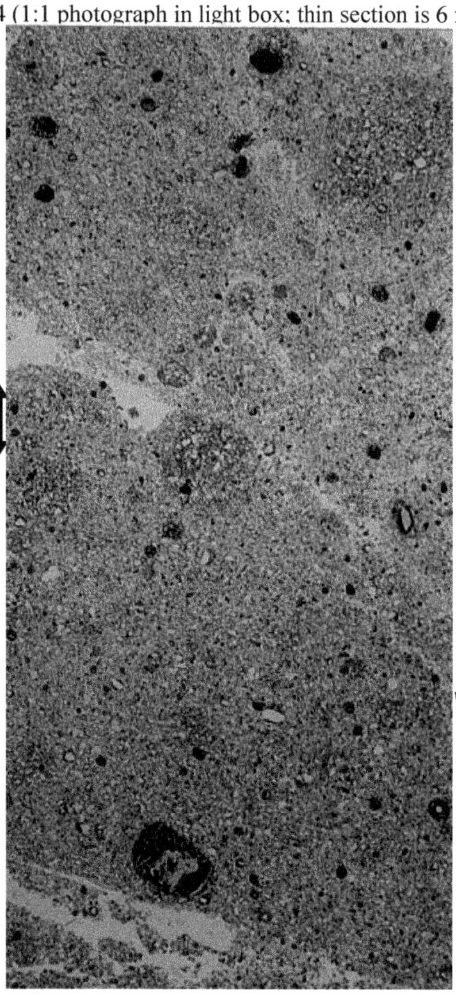

Higher resolution showing basal infilling of fine fraction material, and slight difference in porosity between fill and material in compaction zone (after Jongerius 1983)

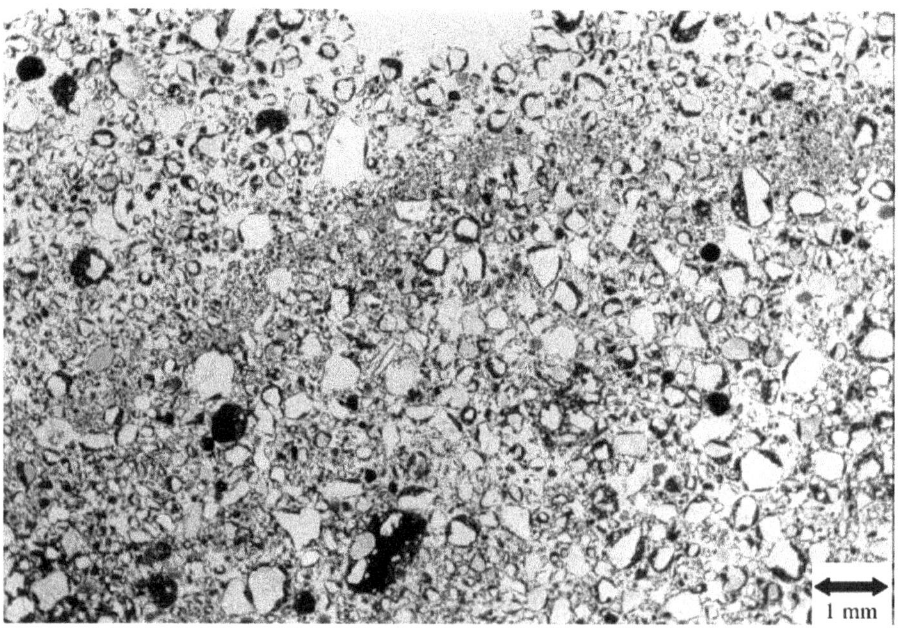

Figure 5 Typical furrow cuts and fills created by the Donneruplund ard at Silsoe soil bin, also showing distortion at base and 'rolled' aggregates in the immediately underlying zone.

Table 11 Summary of microscopic characteristics of ard tilling features and horizons in the Silsoe soil bin

Horizon (depth)	Untilled (SSB1-3, SSB6)	Zone and depth	Tilled
H1 (0 to 4-6cm)	Mixed apedal single grain, single grain as nucleic nodules, granular, occasional subangular blocky-prismatic, large round 'rolled' aggregates, intergrain microaggregates. Porosity 20-30%, packing pores, interpedal and fine intrapedal channels.	**Ridge (5-10cm above surface)**	Large subangular-angular blocky and rounded aggregates (1-3cm or larger) occur at the ridge surface (0-2cm depth), of internal single grain with intergrain microaggregate structure, 20-35% porosity (packing pores). These are underlain by zones of smaller subangular blocky and rounded aggregates (1-2cm or smaller) and, on the ridge side, by a zone of mixed fine aggregates and disaggregated particles and silt. This zone follows the line of the ridge, and is similar in organisation to that seen in the basal fill of furrows.
H1-H2 boundary (4-6cm)	Horizontal void space – planer voids (1000-2000μm wide) and conjoined packing pores, or line of vughs (200-500μm), sometimes partially infilled.	**Ridge-untilled surface boundary (0cm)**	An obvious structural and density boundary, with overlying loose aggregates underlain by undisturbed material of H1 horizon (see 'Untilled' column to left). Diagonally-oriented planar voids (500-2000μm wide), or vughs (700-1000μm), occasionally infilled with fine fraction.
H2 (4-6 to 9-11cm)	Mixed apedal granular, poorly developed subangular blocky-prismatic, occasional large 'rolled' aggregates and intergrain microaggregates. Porosity 15-25%, packing pores, interpedal and intrapedal channels.	**Furrow upper and main fills (0 to 6-9cm)**	After a zone of both small and larger angular-subangular blocky and 'rolled' aggregates at the surface (0 to 1-3cm), with 30-50% porosity (packing pores), the main fill is apedal with disaggregated material, mixed with occasional 2-4cm sized angular and subangular clods, with 20-30% porosity (packing pores).
H2-H3 boundary (9-11cm)	As H1-H2 boundary – linear void space becomes thinner with depth (300-1000μm).	**Furrow basal fill (6-7 or 8-9cm)**	Line of fine components (very fine sand, silt) and fine aggregates following cut shape – aggregate disruption and primary fine fraction deposition. Structure is apedal with very fine rounded aggregates, 20-30% porosity (packing pores).
H3 (9-11 to 14cm)	Apedal granular with rare prismatic or subangular blocky peds, occasional large 'rolled' aggregates and intergrain microaggregates. Porosity 20% as in H2.	**Furrow cut (7-7.5 or 9-10cm)**	A stepped profile is seen when using the arrow-shaped share; cut is a line of interconnected packing pores, planar voids and vugh lines (300-600μm wide), also marked by basal fill of fine components and/or a marked density change from loose (20-35% porosity)-compact (10-20% porosity) zones, and by the presence of very dense 'rolled' aggregates at the boundary between these zones where the arrow-shaped share has cut.
H3-H4 boundary (14cm)	Discontinuous line of rectangular planar voids (400μm wide), or continuous 50μm wide planar voids, or vugh line.	**Furrow compaction/ impact zone (7.5-9 or 10-11cm)**	Planar voids run off from furrow side cut at c. 90°angles, with lifting seen in material outside at sides, related to the impact of downward pressure in the centre of the furrow. This impact appears to be strongest in the surface layers of the soil at the sides, decreasing with depth next to the furrow cut. Dense 'rolled' aggregates and denser zone up to 1 cm thick under furrow (as Jongerius 1983). Porosity 10-20%, mostly packing pores.
H4 (14+cm)	Apedal. Porosity 15-20% packing pores, vughs (<100μm), rare interpedal channels	**Sub-furrow compaction/ impact zone (10-15cm)**	Underlying several furrows, planar voids that echo (run parallel to) part of the furrow base are seen. These occur from 1-4 cm below cut. All other characteristics are as in H4 in untilled (see column to left).

For all: c:f ratio 45-50:50-55 – very coarse sand 2%, coarse sand 5%, medium sand 20%, fine sand 15-20%, very fine sand 15-20%, silt 30-35%, clay <10%; organic components <5%, mostly amorphous subangular/angular fragments of all size ranges and 'punctuations'; mineral components >95% quartz grains, most of the rest feldspar and glauconite, occasional rock fragments of chert and sandstone; chitonic related distribution, occasionally enaulic, some close porphyric in 'rolled' aggregates; groundmass stipple speckled with low birefringence, rare clay zones lining grains, mid-dark brown (PPL), very dark brown (XPL), golden yellow (RL); silt, very fine sand and fine sand coatings on 60-80% of grains, in concentric organisation; horizontal lines of fine components echoing horizon boundaries in some samples.

present macroscopic aggregates, the production of loose zones of fine particles and aggregates, and the deposition of 'lenses' of finer components lining cuts – all through the physical action of the ard itself. The furrow fill samples were taken at the location of powder paint lines, and, as expected, they all show the loss of the surface, upper (4.5-5cm depth) and, rarely, second (9-10cm depth) paint lines. The soil with these paint lines was moved to the ridges and forward by the ard, away from the sampling locations (Lewis 1998).

Furrow cuts in thin section frequently showed the stepped profile expected from the use of the two shares. Immediately underlying the mixed loosened zone of the main furrow fill, almost every thin section with the base of a furrow showed some evidence of physical compaction additional to the background compaction produced by the soil bin processor. In most cases, porosity of the uppermost part of the underlying, untilled layer was lower, both in comparison to the furrow fills and to the deeper horizons (Table 11) (Lewis 1998). In addition, a line of *c.* 1cm-sized 'rolled' aggregates often underlay furrow bases and sides. Such a line of aggregates was frequently encountered under the arrow-shaped share's part of the cut (Fig. 5). The location of a compaction zone within the first centimetre underlying the furrows is in line with findings based on modern tillage implements (Jongerius 1983, 122).

The impact of downwards pressure under the action of the implement was seen in a few samples, where a planar void echoing the ard mark cut is present 1cm below the furrow cut (Fig. 5). In addition, the untilled layers beneath the ard mark have been pushed upwards. The shear planes created around the cut are very important to the location of microscopic features related to secondary processes. The horizon boundary lines created by the soil bin processor frequently showed a tendency to be more compacted (narrower void spaces and/or infilling with grains and fine aggregates) immediately under the furrow base. Some partially compacted planar voids were, however, also created by the soil bin processor itself. In addition to compaction, distortion of lower boundary horizon voids was noted on some occasions. This tended to underlie or be just off to one side of the furrow base (Fig. 6) (Lewis 1998).

The opinion that subsurface compaction is not created by implements such as ards, except possibly as an effect of associated animal trampling (as expressed by Carter and Davidson 1998, 542, citing Macphail *et al.* 1990a), is not supported by these results. Indeed, tillage with *any* implement should be expected to produce compressed zones through downwards- and sideways-moving forces. The added influence of density increase or porosity decrease in the zone of fine material found at ard mark cuts augments subsurface compaction around the depth of tillage. Compaction created under the action of an ard alone is also suggested by the 'gloss' described by Lerche (1994), which lines the entire cut of an implement mark, and cannot be explained as an effect of animal trampling.

Multiple tilling passes produced no further distinctive features regarding furrow fills or cuts, with one possible exception; one sample (SSB19), which was tilled several times, had a pair of parallel planar voids marking the side edge of the cut. No other multiple passes produced similar morphologies (Lewis 1998).

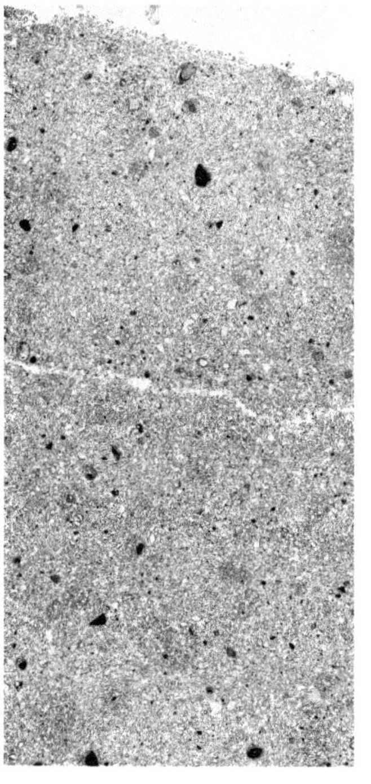

Figure 6 Distortion of material at depth under an ard furrow at the Silsoe Soil Bin (SSB20, scan of thin section; original size 10 x 6cm). The centre of the furrow cut is the surface at the top right; the planar void in the centre of the thin section shows echoing distortion in the same location.

Ridges from Silsoe (SSB7-9)

Ridges were sampled using tubes inserted into the soil from the surface by the soil bin processor. The ridges are characterised by several features, and echo furrow fills in their organisation, being composed of a number of zones. At their very highest points and at the tops of the sides running down into the adjacent furrows, all ridges sampled consist almost solely of 1-3cm or larger subangular-angular and slightly rounded aggregates. The material in the ridge centre is composed of subangular blocky and rounded peds (1-2cm or smaller), mixed with loose particles and fine aggregates, and underlain by the undisturbed soil. All zones are oriented diagonally, following the line of the ridge down into the top of the furrow. This pattern of organisation sees a direct echo to that seen in the ard furrows, with a combination of original soil break-up into larger more angular aggregates (usually seen near the surface), and into smaller aggregates and fine components (usually seen underlying the zone with larger aggregates). Lines of fine materials found at the base of the sides of ridges are similar to those seen in the bases and sides of furrows, except that those in ridges have a slightly higher porosity (Lewis 1998).

The Lejre samples (LEJ6, LEJ8-9)

The overall morphology of the marks from Lejre is similar to that of the Silsoe ard marks. Only the very lowermost part of the fills was sampled for soil micromorphological study (Table 12). The basal furrow fills remaining comprised 0.1-1cm thick layers of single grain with intergrain microaggregate structure, with some fine crumb and subangular blocky peds, and an overall porosity of 10-30% (Lewis 1998). Underlying this in all cases was the untilled layer, which was relatively dense in its upper 1cm (immediately below the furrow cut), again agreeing with evidence from modern cultivation (Jongerius 1983, 122). As at Silsoe, this density boundary appears to be immediately related mainly to compaction of void space under the influence of the traction implement, although a change in texture (increase in fine fraction) from implement action may be involved in creating this zone. No samples external to furrows were taken from Lejre, meaning that the impact of animal trampling cannot be gauged regarding compaction, but the denser zone described here followed the implement cut and was not horizontal to the surface, suggesting it was created by the implement itself.

Linear infillings of clayey silt (silt:clay *c.* 60:40%) associated with the furrow cuts were seen in LEJ8 at the very base of the furrow itself, and at 2mm underneath the part of the furrow created by the arrow-shaped share (Fig. 7), where an infilled shear plane bifurcates, becoming a couple of wavy lines oriented downward underneath the ard mark cut. Also at the base of this furrow was a 4mm thick compact zone. In LEJ9 a similar feature was seen *c.* 3mm under the base of the furrow sampled, and as a void infilling to one side of the furrow (Fig. 8). The infilling features presumably developed within shear planes similar to those created at Silsoe (planar voids echoing and/or just under the cut of the ard mark).

Microaggregates of a size up to that of fine sand comprised a major part of the infilling feature in LEJ8; these underlay a thin clay-rich lens, and overlay a lens of sand. The presence of microaggregates is reminiscent of the lenses created at the base of furrows at Silsoe, and might represent the physical movement of fine aggregates under the action of tilling, although aggregation of fine material *in situ* is also probable at Lejre; in addition, cracking of the infilling is seen. Size sorting of particles was seen in one part of the basal fill in profile LEJ8/1. A fine layer (<200μm thick) of very fine sand particles overlay the silt-clay infilling (Lewis 1998). Such particle sorting was not created through ard disturbance in the soil bin, and it is likely that it relates to the short period of exposure to the elements at this site (*i.e.* to water action), and the cracking seen to wet-dry processes. The creation of these infillings of fine fraction materials at the bases of furrows at Lejre can be seen to replicate Gebhardt's (1990) results with an ard at Hambacher Forst. As in Gebhardt's case, besides these infillings, no other 'dusty' or 'dirty' clay infillings or coatings were clearly created either in the groundmass or in pore space at the base of tillage, despite the fact that an apparently sufficient amount of clay was present. All ard tillage experiments conducted to date for soil micromorphological study have been extremely limited in duration, possibly reducing the opportunity of creating such features, although clay illuviation is known to occur quite quickly at times (such as with the clay component of the Lejre lenses), even in less favourable conditions for dispersion (e.g. Aguilar *et al.* 1983).

One experimental ard mark was not noted in the field, but identified in thin section only (LEJ6), from a plot tilled previously with a mouldboard plough, and then with an ard and left fallow. This had a shallow U-shaped profile and characteristic upper and lower fills as described above (Table 12), along with shear planes and some compaction in the underlying material. This sample showed a line of flattened blocky peds outlining the ard mark cut shape (Fig. 9). The similar location of these and the 'rolled' aggregates at furrow cuts at Silsoe suggests that here, too, these may mark the influence of the implement. This mark was similar to those created during the oxen training runs at Lejre, and its depth in this modern soil (3cm) further supports the suggestion that it was created by the ard rather than the earlier mouldboard ploughing carried out on the experimental site.

Figure 7 Infilling under the base of the furrow in LEJ8 (PPL at top, XPL at bottom).

Table 12 Summary of microscopic characteristics of furrow samples LEJ 8 and LEJ 9, and the ard mark furrow found in LEJ 6

Depth (cm)	**Subhorizon/ Zone**	**LEJ 8** (Ard mark created during oxen training runs)	**LEJ 9** (Ard mark created during oxen training runs)	**LEJ 6** (Field 1 – plot broken with modern plough in 1993, ard tilled twice, then left fallow)
0-2	Upper fill	*Not sampled*	*Not sampled*	Granular, subangular blocky; porosity: 20-40%, packing pores; c:f – 45:55; 15% organics, coarse and fine
2-3	Lower fill	Single grain with intergrain microaggregate; porosity: 10-15, vughs, packing pores c:f – 35:65; 10-15% organics, coarse, fine fragments, some iron replaced; dirty clay and silt lens at base	Single grain with intergrain microaggregate, fine crumb and subangular blocky (1 000 µm); porosity: 20-30%, packing pores, vughs; c:f – 30:70; <10% organics, fine fragments, some coarse	50-200 µm subrounded microaggregates, 2-5 000 µm subrounded aggregates at base; porosity: 40%, packing pores; c:f – 35:70; 15% organics, as above, some large roots; rare iron mottling
3-4	Cut	Silt pan with microaggregates, overlain by lens of very fine sand, texture and density change, void line (discontinuous)	Stepped profile	Line of interconnected packing pores, vughs and planar voids, infilled with microaggregates and fines; shallow, flat U-shaped cut
4-5	Underlying (1 cm)	Vughy to poorly developed subangular blocky; porosity: 10-20%, vughs, channels; c:f – 40:60; 10-20% organics, as above with some calcium carbonate replacement	Intergrain microaggregate, one 2 cm size subangular blocky aggregate; porosity: 10%, vughs; c:f – 45:55; 10-15% organics, as above; silt pan as in *LEJ 8*, but not clearly with microaggregates, 3 mm under furrow base and at side of furrow	Well-developed angular to subangular blocky, elongated peds along edge of cut; porosity: 10%, elongated and rounded vughs, cracks; c:f – 25:75; 15% organics, as above, also horizontally-oriented medium-coarse remains in peds
5 +	Underlying (>1 cm)	Structure as above; porosity: 20%, vughs, channels; c:f –40:60;organics as above :	Intergrain microaggregate; porosity: 15-20%, channels, vughs; c:f – as above; 15% organics, as above	Poorly-developed subangular blocky and granular; porosity: 20%, vughs, channels, cracks; c:f - 40:60; lenses of silty clay, increasing with depth
	Possibly related features at depth			At 15 cm – infillings, possibly akin to weakly-developed agricutan-type features

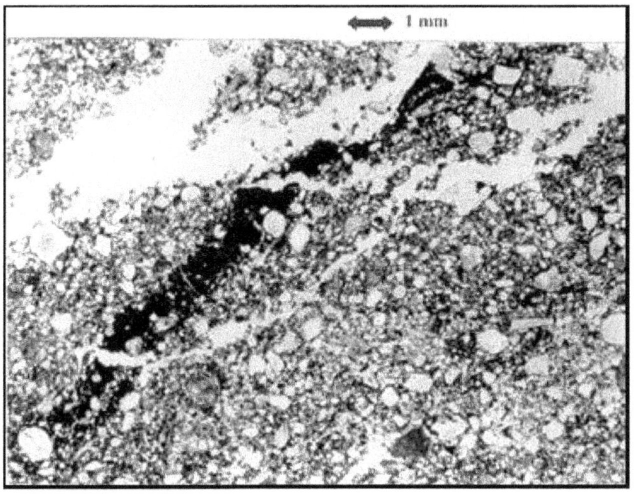

Figure 8 Infilling at the side of a furrow created by the Donneruplund ard replica during oxen training runs at Lejre Historisk-Arkæologisk Forsøgscenter (Sample LEJ 9, PPL). The feature cut is the diagonal planar void on the upper left.

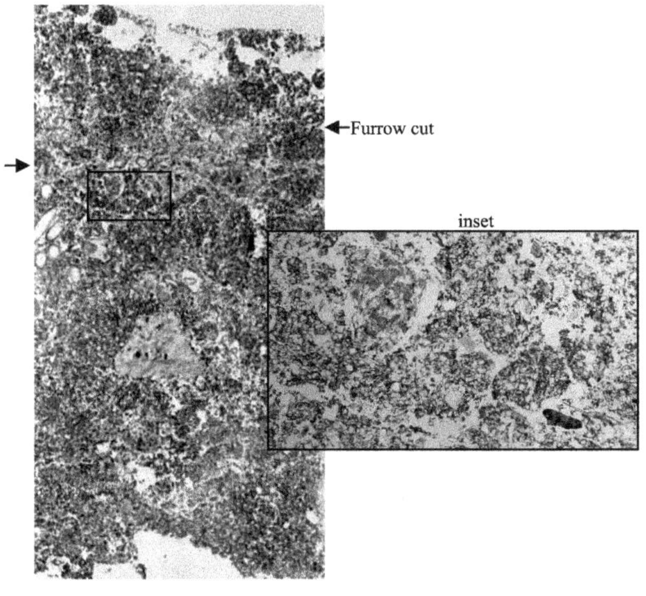

Figure 9 Line of flattened blocky aggregates in LEJ6 marking an ard mark cut (shallow U-shape). (1:1 scan of 6 x 10 cm thin section)

Summary

In the soil bin the 'topsoil' level reacted as predicted through models and as previously described in the literature. The soil was effectively broken up into clods of varying size, creating the typical organisation expected (large, dense angular-subangular blocky peds mixed with zones of small blocky aggregates and disaggregated areas). Microscopically, there are several features which are of interest in regard to possible archaeological identification (Fig. 10, Table 13):

- zones of primary deposition of fine fraction material; in the soil bin, ard tilling leads to the creation of a thin layer of silt and sand grains and fine microaggregates at the base of ard mark cuts. This is a primary depositional feature related to the disruption of clods and soil movement through the action of the ard. The creation of a textural and density boundary at this point of the profile may be expected to play a role in later secondary panning characteristics, and possibly in soil horizon creation in the field.
- external to the furrow, stress-induced fracture or shear planes and voids echoing ard mark bases occur. These are planar voids running either diagonally (at right angles to the sides of the cut) or horizontally (echoing the base of the cut) around the cut. These appear to be created by the combined downward and sideways pressure of the implement. In addition, the layers next to the side of the furrow cut appear to be forced upward. The void spaces created are distinct and strongly-expressed. While it is uncertain how long the voids themselves would survive in the field, they may relatively quickly infill with secondary translocated fine fraction materials, creating diagonally- and horizontally-running linear infillings. There are archaeological examples that demonstrate that such tillage microfeatures do survive archaeologically (e.g. Macphail *et al.* 2004; Lewis 2002; 1998). The infilling of these voids is a secondary depositional feature.
- underneath the tilled zone, deformation occurs, apparently created by the downward pressure of the traction implement. This was expressed as a concave 'dip' in the underlying horizon-marking planar voids at Silsoe, which were set up to be horizontal and straight before tillage. Convex planar voids were also seen, connecting the horizon-marking voids and others echoing these. If such void spaces were to infill under the influence of secondary processes, these might also be expected to survive in some archaeological cases.
- density changes occur at the boundary of the tillage feature. While not particularly strongly-expressed in the soil bin, overall porosity was seen to decrease suddenly at the boundary (Lewis 1998). In addition, some possible physical compaction or rolling was noted in the aggregates located at the surface of the cut, especially under the impact of the arrow-shaped share. As with the lenses of disaggregated finer components lining furrow bases discussed above, the changes in density under and to the sides of the tilled zone should be expected to play a role in secondary feature creation, and possibly in horizonation in certain soils.
- pedality alters between the furrow fill and cut. Basic ped types did not change in the underlying compact zones, but clods were both created and destroyed within feature fills. At the soil bin, large angular clods were cut by the ard at the surface, and deposited within fills and ridges; these aggregates had no precursor in the original structure of the soil.

Where untilled comparative material is available, such structural alterations might be recognised archaeologically (e.g. where large angular peds exist in unexpected settings, ard tillage may be one possible explanation). In addition, the proportion of large 'rolled' aggregates decreased dramatically within the fill. Although some smaller such aggregates were still present, the larger ones appear to have been moved into the ridges and/or broken up into their component parts. The decrease in occurrence of larger aggregates through tillage has been discussed by other authors (e.g. Besnard *et al.* 1996). At Lejre, one sample showed a line of flattened blocky peds following the cut shape.

- the movement of powder paint lines in the soil bin showed both vertical mixing of soil horizons and lateral displacement (forward and sideways) of aggregates from all depths encountered by the shares. Mixing of surface and subsurface horizon materials has been cited as a micromorphological indicator of ancient tillage (e.g. Macphail 1992b; Currie 1994; Lewis & French 2006; Lewis 1998). Most of these cases involve larger and more angular clods of subsurface horizon materials being pulled up into interpreted Ap horizons. This study suggests that smaller and more rounded aggregates should be pulled up from the basal disturbance zone, while large angular clods are produced in the uppermost zones. It is possible, however, that the larger clods found in fills (*versus* those in the ridges) were created at greater depth – the origin of these aggregates was not observed during the experiments. Smaller, rounded clods were moved the greatest distance (up to 1m forward), while larger clods from the surface were restricted in forward movement (up to 30cm). Both saw similar sideways movement into ridges, although large clods tended to be deposited on the outer side, and at the surfaces. Whether any of these arrangements could survive in archaeological situations would depend on burial and preservation environments.
- ridges show a distinctive organisation of structural components, with a zone of isolated large aggregates overlying a mixed zone of broken-up peds and pre-tillage aggregates, underlain by a zone where soil particles and fine aggregates are deposited. These zones are oriented down the ridge slope. Ridges should see the most post-depositional disturbance of all of the zones discussed. Nevertheless, ridges associated with ard marks are sometimes identified archaeologically (e.g. French & Lewis 2001), and it is possible that the structural organisation typical of these features could survive even in less obvious cases.

On the basis of these experiments, it is possible to propose a general organisational scheme of ard tillage impact zones, and certain basic concepts can be supported. It is clear that there is a zone of compaction

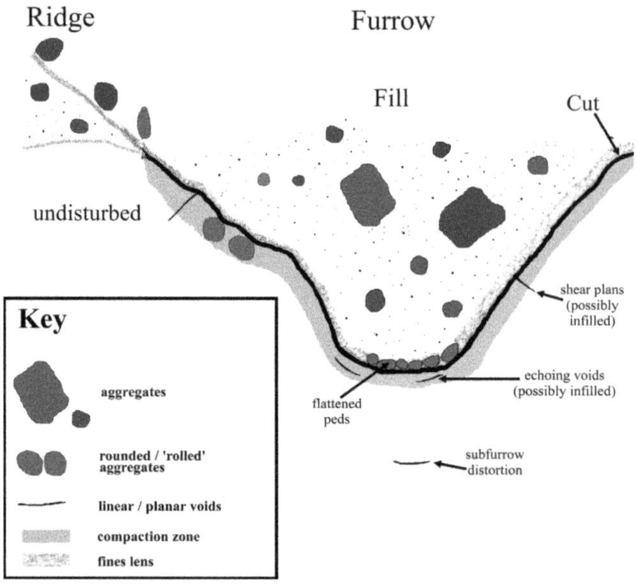

Figure 10 Schematic representation of ard mark feature zones (H. Lewis and C. Stevens, from Lewis 1998)

Table 13 Summary of ard mark feature microcharacteristics

Microcharacteristic	Apparent formation processes
Planar voids in fill	Later implement cuts, shear planes
Planar voids or adjoined packing pores or vughs at cut	Implement entering the earth, removing soil, redepositing soil
Planar voids underlying cut	Shear planes, underneath (echoing) or running at a right angle to the feature cut, presumably related to the downward and forwards (with traction) forces of tilling.
Fragments of surface horizons and subsoil horizons mixed in the main fill	Mixing of topsoil with immediately underlying material
Density increase under and to the sides of cut (decrease in porosity – percentage and size of pores)	Both loosening of tilled soil (fill and overlying) and compaction from implement action
Infillings/pans at/under/to the side of cut	Post-depositional accumulation of translocated fine materials in shear planes/over compacted zones

and increased density associated with the cut itself. This zone may also be associated with 'rolled' or flattened aggregates, and appears to correlate with Jongerius' (1983, 122) 1cm physical compaction zone seen in modern ploughing, and Lerche's (1994) 'gloss'. It is also the location of the accumulation of fine particles and microaggregates moved both through tillage implement action itself (with gravity) and post-depositional translocation. This supports Romans and Robertson's (1983a; 1983b) suggestion that zones of fine fraction accumulation seen archaeologically might relate to base of tillage. It also suggests that 'base of tillage' does not simply equate to some compacted level underneath a cultivated horizon, but to the actual bases and sides of individual implement marks.

Furrow fills and ridges follow a similar pattern of organisation regarding location of types of aggregates. These aggregates may be created at depth, and B, C or lower A horizon clods may be visible in furrow fills as mixing indicators (fabric pedofeatures). Both furrow fills and ridges also include thin layers of fine particles and fine aggregates, which should be expected to have an impact on later profile development, as they play a role in the creation of density boundaries. Finally here, the distinct outline created by the arrow-shaped share was clearly defined both by compaction, rolling and fine material deposition in the soil bin, and should be visible in archaeological situations where marks survive to a depth greater than 5cm.

The results of both sets of experiments suggest that features such as clayey silt infillings/'pans' have a clear morphology that can be directly related to ard tillage, characterised mainly by high proportions of silt, with clay and finer sand-sized components. It is clear that these features are created very quickly and even with minimal disruption and rainfall (in the case of Lejre after only a few summer weeks of shallow tilling a few times per week for less than an hour each time). 'Pans' related to cultivation have been identified micromorphologically in some ancient soils (e.g. Macphail 1987a; 1990a), occurring usually as fragments within or upon individual peds, or characterised by a horizontal zone of decreased porosity and infillings (Bryant & Davidson 1996). The latter zone is different from the features described above, which are discrete textural pedofeatures, not merely zones of increased infilling in general pore space, and not expressed through porosity estimates over a depth. Infillings and pans created at or under tillage feature cuts might be expected to play a role in the expression of further secondary features (e.g. iron or plough pans), if these layers, and the density boundaries they represent, survive for long enough.

One of the most important aspects of the interpretation of ancient tillage using soil micromorphological data is the relationship of microscopic remnant soil features to the processes of tilling. It is vital to understand how soil disturbance can be interpreted and described in such a way as to relate features known to be created by tilling or digging implements, such as ard marks, to features of less certain or possibly mixed origin, such as fabric homogeneity, mixing, panning, or crusts and pore infillings. These results cannot be held to automatically apply to tillage in all types of soils, but through experimental study of ard marks it has been possible to show the relationship of certain characteristic features and organisations to macroscopic disturbance patterns, and to explore physical zones of implement impact on the soil. Although there are many other variations known in cultivated *vs.* non-cultivated soils (see Chapters 2-3), the soil bin experiments were not designed to explore these aspects, but to characterize implement marks and their impact on a sterile soil.

The characteristics identified suggest that the formation of certain types of textural pedofeatures is important in the soil tested and, as reported in the comparative literature (Chapters 2-3), these types of features may survive archaeologically in a variety of environments. Structural characteristics, sometimes related to textural pedofeature formation here, sometimes not, are also clearly produced through experimentation. The following chapters focus further on these two aspects of cultivation through comparison to the remains of other experiments in less controlled conditions.

Chapter 5 Profile and horizon characteristics of soils tilled with ancient tool replicas

Introduction

As part of this study, four experimental sites were assessed in order to examine the general profile and horizon characteristics of soils tilled with reconstructed 'ancient' implements. The aim of this work was to address the issues raised by Carter and Davidson (1998) regarding the difficulties in distinguishing between tillage and other disturbance types, and to assess the model of ancient tillage depth impact presented by Romans and Robertson (1983a; 1983b; see above).

The theme of this volume is how experiments in cultivation can inform us on the characteristics of soils tilled in the past, with a focus on implement impact and soil micromorphological approaches. It is not possible through experimental archaeological work to document long-term changes in soil profiles, such as, for instance, the proposed development of rendzinas from deeper brown earth soils in southern England (e.g. Fisher 1982; French *et al.* 2007; Macphail 1986b; 1987a; 1992a) or of podzolised soils from brown earths in Ireland (e.g. Proudfoot 1958; Conry *et al.* 1996), both of which have been related to prehistoric land clearance, cultivation and land-use intensification through studies of buried soils and erosion history (see e.g. French *et al.* 2007 for a discussion). There are, however, certain characteristics of tilled soil profiles which can be seen experimentally that may relate to these longer-term changes, and which are possible to relate to cultivation events and impacts, thus allowing distinctions to be made between ancient cultivated and grassland areas (e.g. between ancient Ap and Ah horizons). These include many characteristics discussed in Chapter 2, such as structural alterations and textural variations in the profile or even across landscapes, variations in biological activity, and profile mixing, related to the combined influences of tilling and erosion. These processes affect a soil's fundamental characteristics, such as vegetation cover and type, drainage, and bonding properties, and thus have a strong influence on the pedogenetic processes affecting soil profile development.

In this chapter the term 'profile and horizon characteristics' are defined as features of tilled profiles where direct relationship to implement marks is currently unclear. In particular this chapter focuses on Romans and Robertson's (1983a; 1983b) suggestions that depth of tillage impact can be seen in ancient soils and related to implement types, through a variety of characteristics discussed in Chapters 2-4. In particular these characteristics include soil structure and the depth of textural pedofeature concentration. As noted above, these are not the only potential indicators, and are soil-specific; nevertheless, they form a key part of a model outlining how we can identify ancient implement impact on modern soil profiles. These two aspects also fit with the characteristics created in the Silsoe soil bin (Chapter 4), and are linked to the fundamental impact of an implement on the soil, along with post-depositional changes. In this chapter, samples from experimentally tilled brown earth soils in the UK and Denmark are compared to Romans and Robertson's predicted depth of impact with hoes, spades and ards (see Chapter 2), and examined for indications of the implement mark impact zones discussed in Chapter 4.

Background to the experimental sites

The four sites studied are Butser Ancient Farm, Butser Old Demonstration Area, Little Butser and Lejre Historisk-Archæologisk Forsøgscenter, all of which have a history of experiments in ancient agriculture.

There have been three Butser Ancient Farms, all located near Petersfield, Hampshire, UK, and all developed and run by the late Peter Reynolds. Samples were taken for analysis from all three sites – the original farm on Little Butser hill (0.977126°W, 50.982127°N), the Old Demonstration Area (0.583382°W, 50.575735°N) and the recent Butser Ancient Farm site (0.583382°W, 50.575735°N) (Lewis 1998). The sites are located on chalk downland; the soils are rendzinas on chalk, except at the Old Demonstration Area, where there is also calcareous colluvium (Gebhardt 1990).

The Butser cultivation experiments focused on crop production, with fields divided into plots under regimes varying with season of sowing (autumn or spring), crop type and rotation, and manuring or non-manuring. The experiments were not designed to examine tillage specifically, although certain aspects of tillage were discussed by Reynolds (1981). At Little Butser and the Old Demonstration Area, plots tilled with spades, hoes and Iron Age ard replicas were sampled for this study. At the recent Butser Ancient Farm the plots sampled were tilled using hand implements (spades, hoes and forks); these plots had been under Ancient Farm management for four years by the beginning of this project. The first two years of management entailed 'cleaning' the soil of modern fertilisers, by leaving the fields fallow (and untilled). Prior to this, the land was tilled under a modern farming system. This meant that all tilled samples from this site were 'contaminated' by modern processes (P. Reynolds, 1993, pers. comm.).

From the modern Butser Ancient Farm, it was hoped to gather information from ongoing low-impact tillage practices, and from the two earlier sites the aim was to collect comparative material demonstrating post-depositional changes after reconstructed Iron Age cultivation activities had been abandoned to pasture. The Old Demonstration Area was, in addition, one of Gebhardt's (1990) experimental sites, so there was a background soil micromorphological database for comparison from that site.

At Lejre Historisk-Archæologisk Forsøgscenter (11.5636°E, 55.3623°N) experiments focusing on aspects of crop production included a series of tillage experiments that was carried out in the 1960s-70s (Hansen 1969; Lejre Research Center Archives), and more recent experimental cultivation. One plot (Field 3) which had been used to grow flax, tilled for ten years with a Døstrup ard replica (see Glob 1945 for information on this ard type) and left fallow, had been undisturbed since 1994 (M. Rasmussen, 1995, pers. comm.). This plot was sampled during two consecutive summers. In addition, an area in another field (Field 1) had been tilled with the ard once, with six small plots being tilled twice (see Lewis 1998 for field locations). These had been left fallow since the spring of 1995, one year before being first sampled for this project. It was later discovered that the field in which these ard-tilled plots were located had been broken using a mouldboard plough before the ard experiments were carried out, except for one area which had been broken using a tractor-pulled ard (M. Rasmussen, 1995, pers. comm.). This means that the impact of modern cultivation cannot be ruled out regarding the tillage indicators identified, except in the case of the ard mark previously discussed in Chapter 4.

Soil micromorphology study of the experimental sites

Butser Ancient Farm experimental plots

Thirty-three samples were taken from a field at the recent experimental farm, representing 20 profiles (PR1-20) (Table 14) (Lewis 1998). The field was tilled for two years using spades and hoes, and was subdivided into small plots, each with a different manuring and crop rotation regime. Comparative samples from 3 profiles were taken from a pasture field uphill from the tilled plots (FI1-3). This field is thought to have been pasture for a very long time – at least during recent memory (P. Reynolds 1994, pers. comm.), and at the time of sampling it was used for grazing sheep. A background profile located at the edge of the farm was also sampled (Bsamp1/1-3). This was grass-covered and untilled for at least the life of the Ancient Farm, but had been tilled previously with modern machinery. Four additional profiles from one plot (Plot 3) were sampled (F31-4). This plot was cultivated with a metal fork, due to the impact of poor weather on soil moisture. Samples were taken for comparison with the spade- and hoe-tilled samples, and as a record of land use on the plot. Schematic profiles and full profile and thin section descriptions can be found in Lewis (1998).

Background samples – tilled (PR1-2, Bsamp1/1-3)

Profiles PR1-2 were sampled in January 1994, as comparatives to the spade- and hoe-tilled profiles from Plot 3. They are known to have been tilled using modern machinery, and were also left fallow (untilled) for two years before sampling, but were kept relatively clear of covering plants.

The most important characteristics of the profiles in regard to interpretation of the rest of the site are the depths of the subhorizon boundaries and discrete textural pedofeature accumulation zones (Table 15). Several subhorizons were distinguished in all three profiles based on structural and textural changes, and alterations in inclusions. The profiles are subangular blocky to crumb and granular in structure, and in general they have a more well-developed blocky structure with depth. The upper 12cm (H2) of profile Bsamp are more apedal, related to lower porosity and a slightly increased proportion of fine fraction. In this subhorizon there is also a relatively large number (compared to the rest of the profile) of *c.* 2cm-sized chalk rock fragments. These characteristics may reflect the action of modern tillage (subsoil material brought up from depth into the topsoil, a generally less-developed pedality/more apedal nature in the surface horizons).

The more dense nature of this upper subhorizon probably also reflects surface processes. Underlying this is a 3cm thick layer (H3) which possesses textural pedofeatures ('dusty' silt-clay infillings and coatings) unseen in the overlying or underlying subhorizons. All of this together suggests that the 12-15cm depth is a significant boundary in this profile. It is also the approximate depth estimated for spade tillage (Romans & Robertson 1983a; 1983b). This profile is not known to have been tilled with a spade, and whether this boundary zone can be related to the modern tillage that had been carried out is uncertain. The influence of biota can also not be ruled out; perhaps 12cm depth is a rooting zone or the depth limit of some faunal activity here, although no clear indications of either were noted in the field or in thin section.

In samples PR1-2 there are also subhorizon boundaries based on structural changes (pedality and porosity) falling between 12-16.5cm depth (Table 15) (Lewis 1998). These profiles do not possess the textural characteristics mentioned above. Nevertheless, the location of boundary zones at a consistent depth of 12-17cm suggests some impact of modern tilling and/or crops at this depth. This is extremely unfortunate regarding the interpretation of the experimental tilling samples, as this 'background noise' interferes with the identification of features related to predicted depth of spade feature characteristics, assuming that this implement type has an impact zone which should be marked around 12-16cm below the surface (after Romans & Robertson 1983a; 1983b).

The Bsamp profile thin sections include several fabric pedofeatures (Table 15): zones of lower horizon materials are found throughout the profile. These include rounded zones with a reticulate-striated clay-rich fabric (typical of Bt horizons), which are relatively well-integrated with the main groundmass fabric, and zones consisting of organic-stained chalky calcitic fabrics, which are more discrete, and sometimes found in chambers. No B horizon was distinguished in the field (if it still exists, *i.e.* has not been totally integrated into the Ap horizon, it is more than 31cm depth from the surface). The movement of such materials up into the A horizon may be attributed to a combination of faunal mixing, root activity, and, especially in the case of the rounded clay-rich fabric zones, prior tilling activity. A possible surface crust fragment was found at 6cm depth in PR2; this may have been reworked by tilling or other disturbance mechanisms.

In Bsamp, the lack of a well-developed porosity pattern distinctive of grass cover is undoubtedly due to the fact that arable abandonment occurred not long before sampling – the rooting system of the ground cover had not yet had time to strongly influence the structure of the uppermost soil layers. There was also no distinct litter layer in PR1 and PR2.

Background samples – Field I pasture (FI1-3)

A second set of background samples was taken from a field adjacent to the tilled plots, in which sheep were grazed. Samples were taken from three profiles located progressively further uphill from the farmed plots (Table 16). Profile 1 (FI1/1 upper, 1/1 lower, 1/2) was located at the base of the slope, Profile 2 (FI2/1, 2/2) at about halfway up the slope, and Profile 3 (FI3/1, 3/2) at the crest of the hill (see Lewis 1998 for further information).

The samples have a relatively well-developed subangular blocky to crumb and granular structure, reflecting grass cover and a relatively high level of faunal activity. In comparison to those tilled and left fallow (discussed above), the pasture soil is more stained with amorphous organic matter, and shows an increase in number of chalk fragments with depth, rather than in the uppermost part of the profile. No major variation with topographic location was noted, except for a decrease in profile depth going uphill (Lewis 1998), which is expected for a slope, especially where there has been agricultural land use.

As noted in the fallow samples (above), subhorizon boundaries in the pasture samples were seen to coincide with the predicted impact depth of spade tillage (after Romans & Robertson 1983a); in this case there is a change from subangular blocky to predominantly crumb and granular structure (Tables 16 and 17). Again, with no known spade digging on record at this location, it is unclear if the structural changes seen at this depth are created by tillage, or by rooting or faunal activity; the latter seems likely. There were no textural pedofeatures seen in the pasture soils at this depth, unlike in some of the fallow samples.

Surface crusting was seen in the profile from the crest of the hill; this was unexpected (no bare area was noted in the field), but may reflect seasonal impact or possibly increased trampling by animals in this area. Similar crusts were seen in some of the tilled *PR* samples from Plot 3 (see below).

Table 14 List of soil micromorphology samples from Butser Ancient Farm

Sample No.	Description
PR 1/1-2, 2/1-2	Fallow since 1992, non-manured. All PR samples were taken in January 1994.
PR 3/1-2, 4/1-2	Cereal/legume rotation since 1992, non-manured, spade-tilled, hoed
PR 5/1-2	Spring-sown emmer since 1993, non-manured, spade-tilled, hoed
PR 6/1-2	Spring-sown emmer since 1993, manured-non-manured boundary, spade- tilled, hoed.
PR 7/1-2	Spring-sown emmer since 1993, manured, spade-tilled, hoed.
PR 8/1-2, 12/1	Plot 3 - spring-sown emmer since 1992, non-manured, spade-tilled, hoed.
PR 9/2, 10/1-2, 11/1	Plot 3 - spring-sown emmer since 1992, manured, spade tilled, hoed; *PR 9/1* was disturbed in transit.
PR 13/1-2, 14/1-2, 15/1, 16/1-2	Autumn-sown emmer since 1992, non-manured, spade-tilled, hoed.
PR 17/1-2, 18/1, 19/1, 20/1-2	Autumn-sown emmer since 1992, manured, spade-tilled, hoed.
BSamp 1/1-3	Background, from corner of grassy area near PR field, untilled since Butser began, previously tilled with modern equipment, taken in March 1994.
FI 1/1,1/2	Soay sheep pasture, base of slope adjacent to PR field. All FI samples were taken in January 1994.
FI 2/1, 2/2	Soay sheep pasture, halfway up slope adjacent to PR field.
FI 3/1, 3/2	Soay sheep pasture, top of slope adjacent to PR field.
F3 1/1-2, 2/1-2	Plot 3; spring-sown emmer since 1992, manured, spade-tilled, hoed. All F3 samples were taken in April 1994.
F3 3/1-2, 4/1-3	Plot 3 - spring-sown emmer since 1992, non-manured, spade tilling first year, fork tilling second year, hoed.

Table 15 Summary of microscopic characteristics of background and control samples from Butser Ancient Farm (Bsamp, PR1-2); cleared and left fallow (untilled) since 1992, non-manured since Butser began

Profile	Subhorizon	Depth (cm)	Microstructure	Porosity (%)	c:f ratio	Groundmass	Organic components	Textural pedofeatures	Fabric pedofeatures
Bsamp L/Ap	H1	0-1	granular, crumb	15-25%: packing pores, planar voids, channels, vughs	10:90	stipple-speckled	35-35%: coarse tissue fragments, fine amorphous fragments	none noted	in all, but especially H3 - zones of FP1 and FP2 (see Appendix III.ii).
Ap	H2	1-12	apedal crack, weakly developed subangular blocky	10%: cracks, channels, vughs	15:85	stipple-speckled	10-25%: fine fragments, occasional coarse as in H1	none noted	
	H3	12-15	well-developed subangular blocky, granular	30%: cracks, channels, vughs	20:80	stipple-speckled	15-20%: as in H2	silt-clay void coatings, dense incomplete clay-rich infillings	
	H4	15-24	crumb and granular, subangular blocky	20-25%: cracks, vughs, channels	20:80	stipple-speckled	as H3	none noted	
PR1 Ap	H1	0-4	irregular subangular blocky, crumb and granular	20-30: cracks, channels, vughs	15:85	stipple-speckled	10%: fine fragments, 'punctuations'	none noted	H1-H2 horizons: FP1 inclusions (see Appendix III.ii) H2 - also FP2 zone
	H2	4-15	well-developed subangular blocky, apedal crack, crumb and granular, vermicular	15-20: cracks, channels, vughs	15:85	stipple-speckled	as H1	clay/fine silt coatings, channel infillings	
	H3	15-25	subangular blocky, vermicular, crumb and granular	15-30: cracks, vughs, channels, packing pores	10-90	stipple-speckled	as H1	clay/fine silt zones, possibly vermiform-related	FP3 zone
PR2 Ap	H1	0-14	crumb and granular, poorly-developed subangular blocky, apedal crack and vugh, vermicular	15-20%: vughs, packing pores, cracks, channels	15:85	stipple-speckled	10%: as in PR 1	6 cm depth - crust fragment	all horizons: FP1, FP3 and FP4 zones (see Appendix III.ii)
	H2	14-16.5	crumb and granular, subangular blocky	30%: channels, cracks, vughs	15:85	stipple-speckled	as H1	none noted	
	H3	16.5-22	well-developed angular to subangular blocky	15-25%: cracks, vughs	15:85	stipple-speckled	as H1	none noted	

Table 16 Summary of microscopic characteristics of background and control samples from Butser Ancient Farm (FI 1-3); pasture

Profile	Subhorizon	Depth (cm)	Microstructure	Porosity	c:f ratio	Groundmass	Organic components	Textural pedofeatures	Fabric pedofeatures
FI 1 L	H1	0-0.2	crumb and granular, subangular blocky	40-50%: packing pores	30:70	stipple-speckled	50-60%: litter, 'punctuations', organic staining, rare charcoal	none noted	all horizons: fabric pedofeatures FP1, FP3,
Ah	H2	0.2-8	well-developed subangular blocky	15-20%: cracks, channels and vughs	35:65	crystallitic	20-30%: as H1, some iron replaced	occasional clay infillings, sandy zones related to earthworm sorting	FP4 (see Appendix III.ii)
	H3	8-12	well-developed subangular blocky, crumb and granular, vermicular	10%: as H2	35:65	crystallitic and stipple-speckled	as H2	as in H2	associated with rooting,
	H4	12-19	crumb and granular, apedal, moderately-developed subangular blocky	15-20%: as H2	35:65	crystallitic	20-25%: as H2	none noted	earthworm sorting
	H5	19-24	moderately-developed subangular blocky, crumb and granular	15-20%: as H2, but one spongy zone	35:65	crystallitic	as H4	none noted	
FI 2 L/Ah	H1	0-7	moderate to well-developed subangular blocky, crumb and granular	20-30%: cracks and channels	30:70	crystallitic and stipple-speckled	10-30%: fine fragments, litter zones, as in *FI 1 H1*	none noted	both horizons: zones of PF1 (see Appendix III.ii)
Ah	H2	7-18	crumb and granular and subangular blocky, spongy with depth	30-35%: as H1, also vughs and packing pores	30:70	as H1	10-20%: fine fragments, rare charcoal	none noted	
FI 3 L	H1	0-1	Crumb and granular	70%: packing pores	50:50	stipple-speckled	70-80%: litter, as in *FI 1 H1*	none noted	none noted
Ah	H2	1-10	well-moderately developed subangular blocky (apedal where surface crust)	20%: cracks, channels, vughs	30:70	crystallitic	10%: fine fragments, some iron replacement, some coarse structural material	surface crust	none noted
	H3	10-13	crumb and granular, irregular blocky	10-15%: as H2	30:70	as H2	10-15%: as in H2, also root remains	sandy zones related to earthworm sorting	zones of FP3 and FP4
	H4	13-17	moderately-developed subangular blocky, prismatic, crumb and granular	20%: as H2	25:75	as H2	10-15%: 'punctuations', rare charcoal, coarse material with cellular structure	rare slightly dusty clay zones in groundmass, sandy zones related to earthworm sorting	(see Appendix III.ii)

Spade-tilled samples – Field 3, Plots 1-4 (PR3-20)

Samples were taken from twenty profiles (Plot 3; Lewis 1998), which had seen spade tilling for 2 years prior. In several cases structural boundaries and textural features (or fragments of these) correspond in location with the expected depth for textural features created by hoe tilling (4-5cm) and spade tilling (10-15cm) (after Romans & Robertson 1983a; Gebhardt 1990) (Tables 18-22). It might be expected that the changes in structure and porosity seen at these boundaries could be related to potential zones in which fine fraction-based microfeatures associated with implement disturbance zone depth could accumulate in appropriate soils, as discussed in relation to ard marks in Chapter 4. However, the background samples, none of which were reported to have been tilled with these implements, also show boundaries and rare textural feature fragments at these depths. And the suggestion that the 12-16cm depth zone marked in many background samples from Butser Ancient Farm is related to rooting and/or faunal activity is supported by the evidence of some of the tilled samples (Lewis 1998). PR3, for example, had a boundary at 13cm depth marked by a stone line (often associated with earthworm activity) and roots replaced by calcium carbonate (Table 18). PR10 has large channels created by biota in a zone at 16-17cm depth (Table 20). Sample F3 1, discussed below, has a rooting zone at 8.5-13cm depth (Table 22). Although several structural boundaries and textural pedofeatures in the tilled samples do correspond with the depths of predicted impact (after Romans and Robertson 1983a; 1983b), and may relate to tillage depth, this is clearly not a straightforward indicator of implement type used, instead reflecting a number of processes acting on the soil.

The profiles show major characteristics associated with biologically-active and tilled soils (Tables 18-21), and can be considered typical of tilled rendzinas in terms of structure, showing three main types of profile:

- profiles with dominantly crumb and granular structure, becoming blockier or sometimes apedal with depth (PR4, PR6, PR8-12, PR20). Most of these are spring-sown samples.
- profiles which are predominantly subangular blocky, with zones of crumb and granular structure (PR5, PR7, PR15-16, PR19).
- profiles with a more apedal nature, especially in surface (sub)horizons, and less well-developed blocky structure and crumb and granular with depth (PR3, PR13-14, PR17-18). Most of these are autumn-sown samples, with the apedal surface structure being related to surface sealing.

Besides these patterns, there are no distinctions apparent regarding any other variables, including manuring, in the tilled profile structures. All samples show at least some vermicular structural elements. Laminar peds were seen in the surface horizon of PR20; these have been related to modern tillage (Fitzpatrick 1984, 126) and are associated with compaction. Laminar peds are also present in one of the fork-tilled samples (see below).

Table 17 Subhorizon boundaries and textural pedofeatures from Butser Ancient Farm background samples, compared to predicted depth of impact of hoe and spade tillage (after Romans and Robertson 1983a & b; Gebhardt 1990)

	Fallow for 2 years	Cleared, fallow for 2 years		Pasture			
Depth (cm)	Bsamp	PR 1	PR 2	F1 1	F1 2	F1 3	
0-2	L	A(p)	A(p)	L	Ah	L	
2-4	A(p)			Ah		crust	
4-6		A(p)	crust			Ah	Predicted impact depth of hoe tillage
6-8							
8-10				Ah	Ah		
10-12						Ah	Range of predicted impact depth of spade tillage
12-14	text pf			Ah			
14-16	A(p)		A(p)			Ah	
16-18	A(p)	A(p)					
18-20			A(p)				
20-22				Ah			
22-24							
24-26							
26-28							
28-30							
30-32							
32-34							
34-36							

KEY

- - - - subhorizon boundaries
crust - includes crust fragments
text pf - infillings and coatings
______ - base of excavation

L litter Ap – tilled topsoil

The more compact nature of the autumn-sown samples could be explained by the more recent tilling of these plots before sampling in the following January, with subsequent translocation of fine fraction, creation of crusts, and weathering over the autumn and winter. The mainly subangular blocky profiles show a mixed structure reflecting both tilling and faunal (especially earthworm) reworking. The spring-sown emmer (for 2 years) profiles are all mainly crumb and granular, suggesting a slightly stronger influence of soil fauna *versus* tillage in these samples. These variations, which appear in two sets of cases to correspond with season of sowing (and tilling) might form a sort of continuum reflecting length of time since and season of last tilling before sampling. While a strong 'tilled' structure and organisation appears after autumn sowing, either the same strength of tilled structure is not produced after spring sowing, or the amount of biological reworking of these soils is such that after as little as one year (possibly one summer) before sampling, structure has changed to predominantly crumb mainly through earthworm activity. Similar results were reported by Gebhardt (1990) from this site and from Lann Gouh in France (also see Macphail 1990b). This very short period of time for major structural reworking has implications for the identification of ancient tillage from structural indicators in rendzina soils.

There was no crusting seen in the spring-sown profiles. This is somewhat surprising, given the time of year in which they were sampled and that the fields were bare. This suggests that finding crusts (as seen in the autumn-sown profiles) is not necessarily simply a factor of having bare earth conditions in disturbed soils, but also relates to immediate subsequent reworking activities, *i.e.* exactly the same cultivated soil may produce/not produce or retain/not retain crust fragments depending on season of tilling (and sampling). Archaeologically it would be quite difficult to know whether a buried rendzina soil has been bare or not; if spring-sown, evidence of surface features of bare earth may be erased within one season. In this case, based on the strongly crumb and granular nature of these profiles, any crusting features appear to have been reintegrated into the main groundmass of the soil by soil faunal and plant rooting activity during the summer.

Linear infillings/'pans' and crusts are, however, clearly distinguished in those Butser samples in which both occur (Figs. 11-13). In brief, pans generally show a relatively low birefringence, are relatively homogeneous, and have an overall finer texture (fine silt and clay, with organic staining) and denser structure. Crusts at Butser were frequently banded, with fine layers of coarse silt, finer silt with little clay/amorphous material, and fine textured bands, but with some larger components. They were more influenced by coarse silt and sand content and were relatively bright regarding birefringence (Lewis 1998). Most commonly seen were infilling structural crusts (bare coarse silt and fine sand grains in pore spaces), thought to originate from slow erosion of the surface (Valentin & Bresson 1992, 229), and runoff depositional crusts (bedded texturally-sorted layers, which usually form after sealing by structural crusts) (*ibid.* 233; Mücher & de Ploey 1977). Some possible slaking structural crusts were also identified (after Valentin & Bresson 1992, 229-30).

While one crust was seen in a pasture sample (discussed above), no similar silt-clay 'pans'/infillings were observed in the pasture profiles. Although the sample number is small, given the relationship proposed between such microscopic features and tillage features from the Lejre ard mark samples, it is possible that these features seen in the tilled samples at Butser do indeed relate to tilling. One of the linear infillings identified correlates with the zone of expected location of textural features from hoeing, but the other is slightly deeper than that expected from spade tilling (Table 18), and it is possible that the latter, at least, is related to the earlier land use discussed above. The presence of a crust in one of the pasture samples suggests that not only is it difficult to relate such features to tilled soils, but also that the interpretation of these features as indicators of land clearance for cultivation should also be questioned. It seems that minor animal trampling and/or the presence of small bare patches in an otherwise undisturbed plot of land can also produce such features with at least as much regularity as is seen in bare tilled plots (e.g. as compared to the PR samples taken in the same season).

A 'whole soil' coating was also seen in PR19 (see Fig. 2 in Chapter 2). Its position and morphological similarity to parts of the observed surface crusts suggest that it is probably related to surface sealing and is not 'agric' (c.f. Fitzpatrick 1984) in origin.

Groundmass characteristics vary between being crystallitic (PR10-16) and stipple-speckled (all others). This is not apparently related to anything in the tillage or manuring regimes, but appears to be localised in one part of the field. In most profiles there are fabric/depletion pedofeatures with variations in groundmass. Most samples include rounded zones of a reddish (plane polarised light) stipple-speckled fabric which is similar to the main fabric in all ways, but without either the micrite or the coarse components. A second fabric pedofeature type consists of zones of reddish fabric in plane polarised light, which is the same as the main groundmass in cross-polarised light. Zones of both of these fabrics vary in size, and are possibly related to rooting or faunal activity. Finally, there are zones of very calcitic fabrics: one is micritic, occurring in channels or as larger aggregates, while another is chalky material, which is presumably related to chalk weathering (Lewis 1998).

No significant variation was seen microscopically between manured and non-manured plots in terms of structure and organic matter content. Gebhardt (1990) saw major differences in cation exchange capacity, and in micromorphologically-expressed organic matter content and structure between manured and non-manured experimentally tilled soils at Butser Old Demonstration Area. Manured soils at the old site were richest in organic matter and showed a crumb and subangular blocky structure, in addition to having small organo-mineral aggregates. In comparison, non-manured samples were poorer in organics and had a friable small angular-subangular structure (Gebhardt 1990, 45, 63). No such correspondence was seen to exist from the samples at the recent Butser site. Perhaps the immediate survival of such manuring indicators varies with soil type and formation controls, even where soils are quite similar. A few possible manuring indicators were, however, found from the site (Lewis 1998): fragments of what might be dung (J. Heathcote, 1997, pers. comm.) were found in samples PR13-15, all of which were said to be non-manured (P. Reynolds, 1993, pers. comm.). These samples were all located in the same subplot, suggesting either some localised animal presence at some time, or that the recording of this plot as non-manured was a mistake.

Fork-tilled samples - Plot 3 (F31-4)

Samples were also taken from four profiles in Plot 3 (Table 23). This plot was the only one tilled since the previous sampling (discussed above). It was tilled with a metal fork, due to the impact of poor weather on soil moisture. The profile structures are mixed, but predominantly subangular blocky, with areas of crumb and granular, angular blocky and vermicular structure in most samples, again reflecting both tillage and strong reworking by soil fauna. One sample had laminar peds (F34/1), which have been related to cultivation (Fitzpatrick 1984). The fact that they are not dominated by a crumb structure despite being spring-sown (see above) might reflect the season of sampling, if the scenario regarding a continuum of structural change related to season is accepted. These samples were taken in the spring, not long after tilling and sowing, while the spring-sown samples from the previous year had been taken in January. It could be that the lack of a more biotically-active season (summer) before time of sampling has meant that a more tillage-related structure is seen.

Infilling feature at 18cm depth in PR 4/2, PPL (left) and XPL (right); arrow length represents 1 mm

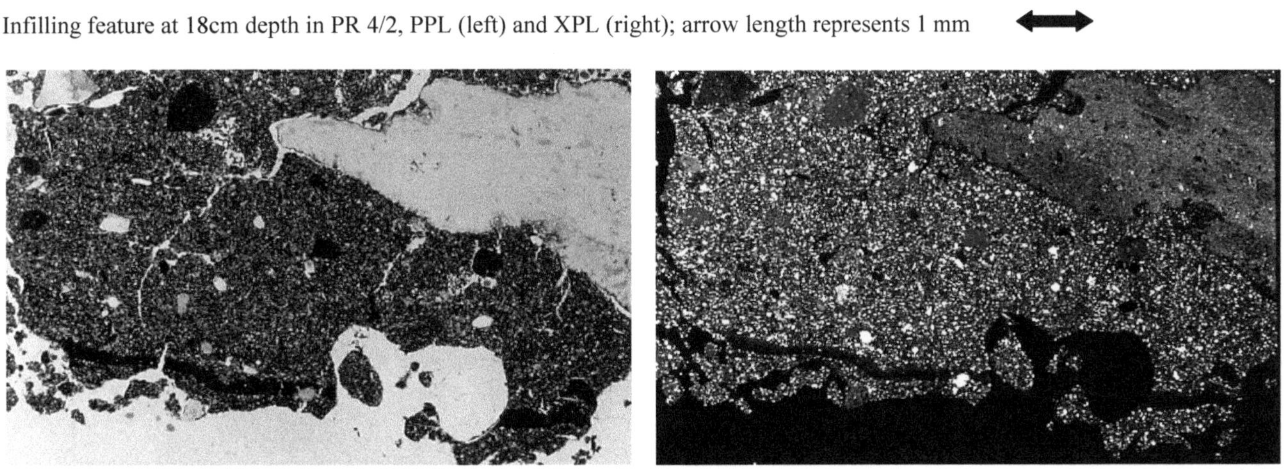

As above, at higher magnification, PPL (left) and XPL (right); arrow length represents 1mm

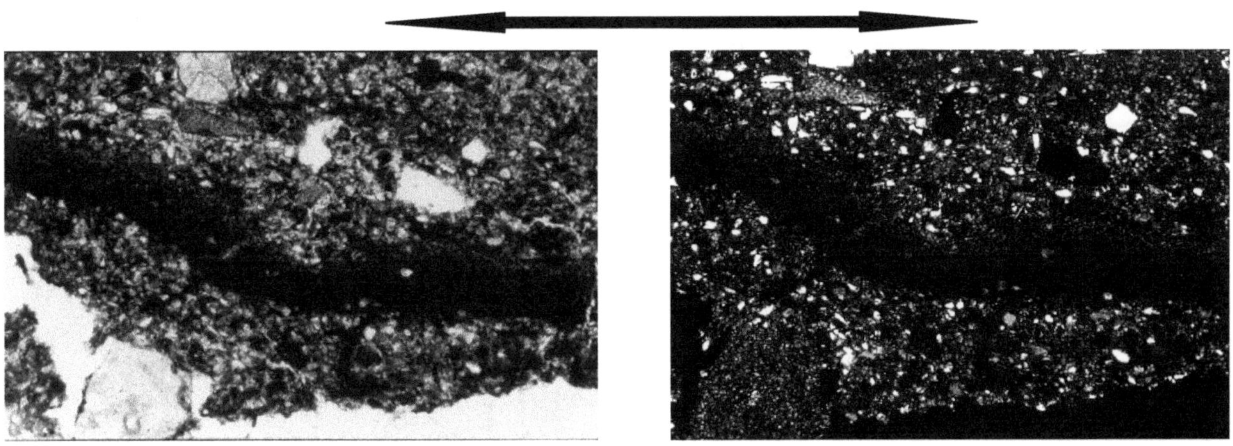

Figure 11 Infilling features related to tillage from Butser Ancient Farm: sample PR 4/2 at 18 cm depth

Depth (cm)	PR5	PR6	PR7	PR3	PR4	PR8	PR9	PR10	PR11	PR12	PR13	PR14	PR15	PR16	PR17	PR18	PR19	PR20	
	Spring-sown emmer, spade-tilled and weeded with hoe for 1 year			Cereal/legume rotation, spade-tilled and hoed for 2 years		Spring-sown emmer, spade-tilled and weeded with hoe for 2 years					Autumn-sown emmer, spade-tilled and weeded with hoe for 2 years								
0-2	Ap	Ap	Ap	Ap	Ap	L	No	Ap	Ap	Ap	Ap	crust	Ap	crust	crust	crust	crust	Ap	
2-4					crust	Ap	sample					Ap			Ap	Ap	text pf		
4-6									'pa n'		Ap						Ap		hoe
6-8		Ap		crust								Ap		Ap		Ap		crust	
8-10																			
10-12	text pf							Ap											
12-14						text pf	Ap				Ap								spade
14-16				Ap	Ap	Ap													
16-18														Ap	text pf				
18-20																			
20-22		Ap		text pf															
22-24																			
24-26					Ap														
26-28																			
28-30																			
30-32																			

KEY

- text pf - textural features at a clearly-defined depth
- crust - surface crust or crust fragment
- pan - panning feature
- - - - - subhorizon boundary
- ——— base of excavation
- L - litter
- Ap - tilled A

Depth (cm)	F3 1	F3 2	F3 3	F3 4	
	Spring-sown emmer, spade-tilled and weeded with hoe for 2 years, fork turned in 3rd year				
0-2	Ap	Ap	Ap	Ap	
2-4					
4-6				Ap	hoe
6-8					
8-10	Ap				
10-12	Ap				
12-14			text pf	Ap	spade
14-16	Ap	Ap	Ap		
16-18				'pan'	
18-20			Ap		
20-22					
22-24					
24-26					
26-28				C	

Shaded areas represent expected base of tilled zones for hoe (4-5 cm) and spade (10-15 cm) according to previous experimental and archaeological findings (Gebhardt 1990; Romans and Robertson 1983a & b).

Crusting in autumn-sown profiles is thought to relate mainly to season of tilling and sampling.

Table 18 Subhorizon boundaries and textural pedofeatures from Butser Ancient Farm tilled samples, compared to predicted depth of impact of hoe and spade tillage)

PR 18/1 Depositional crust

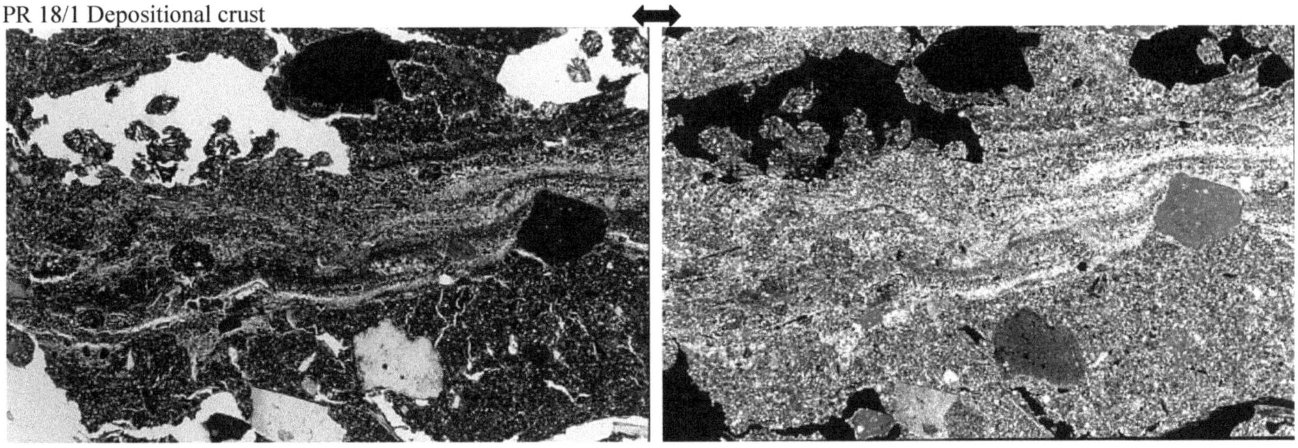

Figure 12 Crusts and surface seals from Butser Ancient Farm: depositional crust in PR 18/1, PPL and XPL; arrow length represents 1mm

PR 19/1 crust and sealing/infilling; arrow length represents 1mm

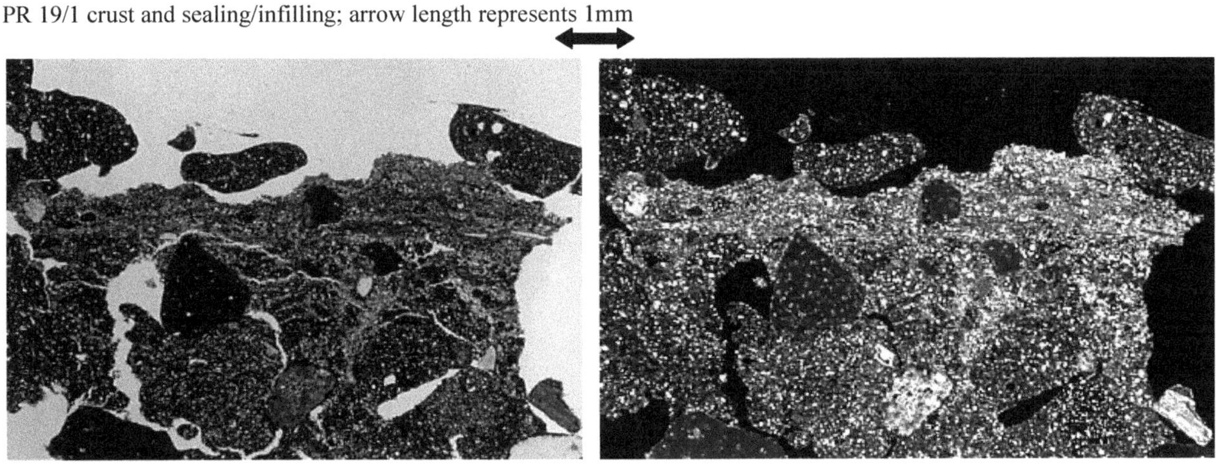

As above, at higher magnification; arrow length represents 1mm

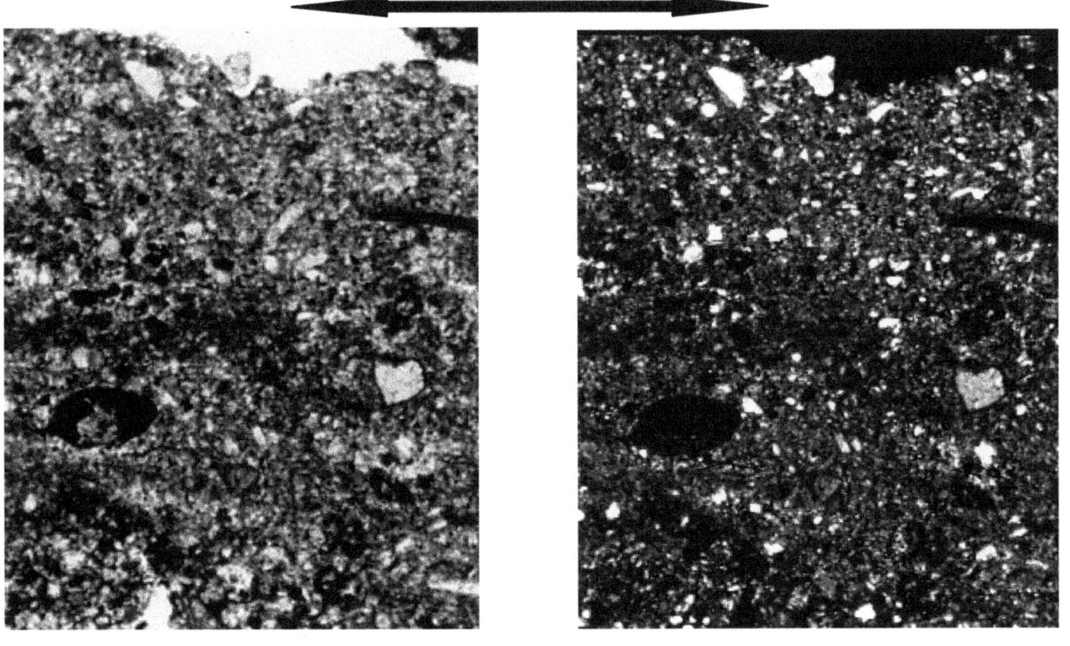

Figure 13 PR 19/1 Crusts and surface seals from Butser Ancient Farm: depositional crust and underlying sealing/infilling of macropores, PPL (left) and XPL (right)

Table 19 Microscopic characteristics of tilled samples from Butser Ancient Farm: cereal/legume rotation for 2 years, non-manured, spade-tilled and hoed

Profile	Subhorizon	Depth (cm)	Microstructure	Porosity	c:f ratio	Groundmass	Organic components	Textural pedofeatures	Fabric pedofeatures
PR3 Ap	H1	0-9	apedal vughy, crumb, well-developed angular blocky	10-20%: vughs, channels, cracks	20:80	stipple-speckled	10%: 'punctuations', medium-sized structural remains	H1 - crust fragment at 7cm depth H1-2 - clay infillings in some pores, rare silty-sandy zones related to vermiforms	all horizons: FP1, FP2 and FP3 zones (see Lewis 1998 Appendix III.ii)
	H2	9-13	crumb, apedal vughy, vermiform	25-35%: packing pores, vughs, cracks, channels	20:80	as H1	as H1		
	H3	13-23	as H2, but some blocky peds	30-40%: as H2	30:70	as H1	as H1	rare dusty clay zones	
PR4 Ap	H1	0-13	crumb, subangular and irregular blocky, vermicular	20-40%: vughs, channels	30:70	stipple-speckled and crystallitic	all horizons: 10-20%: fine particles, some iron- and calcium carbonate- replaced coarser remains	crust fragment at 2.5cm depth, some pendants of silt/sand	H1 - zones with increased micrite
	H2	13-24	crumb, vermicular, angular and subangular blocky	30-40%: channels, cracks, packing pores	25:75	as H1		rare dusty clay zones	all horizons: FP2 zones (see Lewis 1998 Appendix III.ii)
	H3	24-32	crumb, subangular blocky	20-30%: cracks, channels, packing pores, vughs	30:70	as H1		as H3	H2-H3 - FP1, FP2, FP3 & FP4 zones (see Lewis 1998 Appendix III.ii)

Table 20 Microscopic characteristics of tilled samples from Butser Ancient Farm: Spring-sown emmer for one year, non-manured and manured, spade-tilled and hoed

Profile	Subhorizon	Depth (cm)	Microstructure	Porosity	c:f ratio	Groundmass	Organic components	Textural pedofeatures	Fabric pedofeatures
PR5 NM Ap	n/a	0-23	subangular blocky, crumb, vermicular	15-20%: cracks, channels, vughs	20:80	stipple-speckled	<10%: fine fragments, some coarse, rare charcoal	vertical channel infilling (sealing of macropore) at 10-12 cm depth. Zones of fine sand, related to worm sorting	occasional FP1 and FP2 zones (see Lewis 1998 Appendix III.ii)
PR6 NM/M Ap	H1	0-3	crumb, angular-subangular blocky	10-20%: cracks, channels, vughs	30:70	stipple-speckled and crystallitic	15%: fine and medium fragments, coarse cellular fragments, roots, seeds	silt and sand zones related to earthworms	frequent zones of FP1 (see Lewis 1998 Appendix III.ii)
	H2	3-14	subangular blocky, crumb, vermicular	20%: cracks, channels, vughs	35:65	stipple-speckled	as H2	none noted	H2-H4: FP1 & FP4 zones (see Lewis 1998 Appendix III.ii)
	H3	14-20	as H2, but less well-developed	20-25%: cracks and channels, packing pores	as H2	as H2	5-10%: 'punctuations', fine particles, occasional coarser cellular fragments	none noted	
	H4	20-23	as H2, becoming moderately-poorly developed blocky, crumb, apedal crack with depth	15-20%: cracks, channels, vughs	as H2	as H2	as H3	rare void coatings of dusty clay	
PR 7 M Ap	n/a	0-22	well-developed subangular blocky, crumb, vermicular	30%: cracks, channels, vughs	40:60	stipple-speckled	10-20%: fine fragments	silty infilling at 21 cm depth	FP2 & FP3 zones (see Lewis 1998 Appendix III.ii)

Table 21 Microscopic characteristics of tilled samples from Butser Ancient Farm: Spring-sown emmer for two years, non-manured and manured, spade-tilled and hoed

Profile	Subhorizon	Depth (cm)	Microstructure	Porosity	c:f ratio	Groundmass	Organic components	Textural pedofeatures	Fabric pedofeatures
PR8 **NM** L	H1	0-2	crumb	70-80%: packing pores	10:90	stipple-speckled and crystallitic	80%: fresh and decomposing coarse and medium remains	bands of more clay-rich infillings	all horizons: frequent FP3 zones (see Lewis 1998 Appendix III.ii)
Ap	H2	2-12	crumb, subangular blocky, vermicular	40-45%: cracks, channels, vughs	35:65	crystallitic and stipple-speckled	15%: 'punctuations', fine particles, some degraded medium-coarse remains	earthworm sorted silt/sand zones	
	H3	12-20.5	crumb, subangular blocky, vermicular	25-30%: cracks, channels, vughs	as H2	crystallitic	10%: as H2	clay-silt coatings	H3 - FP1, FP3 & FP4 zones
PR9 **M** Ap	n/a	14-22	crumb, subangular blocky	15-20%: channels, vughs	25:75	stipple-speckled and crystallitic	<10-15%: fine fragments, occasional coarser remains	occasional clay/silt coatings, sand/silt sorting (vermiform)	FP1 & FP2 zones (see Lewis 1998 Appendix III.ii)
PR10 **M** Ap	H1	0-9.5	crumb, subangular blocky	15-20%: cracks, vughs, packing pores	40:60	crystallitic	15%: fine fragments, coarser degraded remains, some iron-stained	fine clay/silt coatings, sandy zones related to worms	FP1 & FP3 zones (see Lewis 1998 Appendix III.ii)
	H2	9.5-17	apedal crack, subangular blocky, crumb, vermicular	15-15%: vughs, cracks, channels	35:65	crystallitic	<10%: 'punctuations', fine fragments	none noted	
PR11 **M** Ap	n/a	0-9	crumb, subangular blocky, vermicular	20-35%: packing pores, vughs, channels, cracks	20:80	stipple-speckled and crystallitic	15%: fine fragments, some coarser remains	silt/clay pan at 5 cm depth	FP3 & FP4 zones (see Lewis 1998 Appendix III.ii)
PR12 **NM** Ap	n/a	0-12	crumb, apedal, subangular blocky, vermicular, crumb	20-30%: cracks, packing pores, vughs	25:75	crystallitic	15%: fine fragments, some coarser remains	silt and sand zones related to worm sorting	FP2 & FP3 zones (see Lewis 1998 Appendix III.ii)

Table 22 Microscopic characteristics of tilled samples from Butser Ancient Farm: Autumn-sown emmer for two years, non-manured and manured, spade-tilled and hoed

Profile	Subhorizon	Depth (cm)	Microstructure	Porosity	c:f ratio	Groundmass	Organic components	Textural pedofeatures	Fabric pedofeatures
PR13 NM Ap	H1	0-4	apedal crack, crumb	10-25%: channels, cracks, vughs	40:60	stipple-speckled and crystallitic	10-25%: fine fragments, larger tissue fragments, some iron replacement	channel infilling	all horizons, especially H3 - FP3 zones (see Lewis 1998 Appendix III.ii) - possible dung fragment
	H2	4-11.5	weakly-moderately developed subangular blocky, angular blocky, crumb	as H1	40:60	as H1	As H1	silt capping	
	H3	11.5-19.5	apedal crack and vughy, subangular blocky	15%: vughs, cracks	30:70	as H1	as H1, also rare root remains	sandy zones - earthworm sorting	
PR14 NM Ap	H1	0-5	Angular blocky, apedal crack	15%: cracks, vughs	25:75	crystallitic	10-15%: fine fragments, medium-fine cellular remains, roots	surface crust, silt void coatings, earthworm sorted sand/silt zones	FP2, FP3 & FP4 zones (see Lewis 1998 Appendix III.ii) possible dung fragment
	H2	5-18	crumb, apedal vughy, subangular blocky	20-25%: vughs, cracks	45:55	as H1, becoming chalky	As H1, rare large root fragments	earthworm sorted sand/silt zones	
PR 15 NM Ap	n/a	0-12	subangular blocky, crumb, vermicular, apedal crack	15%: cracks, vughs, channels	30:70	stipple-speckled and crystallitic	10-15%: fine fragments, possible dung inclusion	silty/sandy zones - earthworm sorting	FP3 & FP4 zones (see Lewis 1998 Appendix III.ii)
PR16 NM Ap	H1	0-14.5	well-developed subangular blocky, crumb	15-20%: cracks, vughs, channels	45:55	stipple-speckled and crystallitic	10%: fine fragments, some larger degrading remains, rare charcoal	surface crust, some channel infillings	fabric pedofeatures as interrupted void coating (see Lewis 1998 Appendix III.ii)
	H2	14.5-21	subangular blocky, crumb, crumb, vermicular	20%: cracks, channels, vughs	40:60	as H1	10%: as H1, occasional very coarse remains	none noted	none noted
PR17 M Ap	n/a	0-18	weak-moderate subangular blocky, apedal crack and vughy, crumb, vermicular	10-20%: channels, cracks, vughs	35:65	stipple-speckled and crystallitic	10-15%: fine to coarse fragments, large root remains	surface crust, slightly dusty clay coatings and infillings, earthworm-sorted zones	FP3 &FP4 zones (see Lewis 1998 Appendix III.ii)
PR18 M Ap	H1	0-6	apedal, weak-moderate angular-subangular blocky	20-25%: cracks, channels	15:85	stipple-speckled and crystallitic	20%: frequent coarse remains, roots, charcoal, fine fragments	surface crust, occasional void coatings	FP3 and FP4 zones (see Lewis 1998 Appendix III.ii)
	H2	6-10	apedal crack, irregular subangular blocky	25%: cracks, channels, vughs	20:80	stipple-speckled and crystallitic	10-15%: as H1, also frequent fine and coarse degrading remains	none noted	
PR19 M Ap	n/a	0-10	well-moderate subangular blocky, apedal vughy, crumb	10-20%: channels, vughs, cracks	25:75	stipple-speckled and crystallitic	10-20%: fine-medium fragments, occasional very coarse remains	surface crust, associated laminae at 3 cm depth, zones of silt/sand related to earthworms	FP3 & FP4 zones (see Lewis 1998 Appendix III.ii)
PR20 M Ap	H1	0-9	crumb, subangular blocky, laminar, spongy	20-30%: cracks, vughs	30:70	stipple-speckled and crystallitic	10-20%: decomposing coarse and medium remains, fine fragments	crust fragment at 6.5cm depth, rare very dirty clay coating	FP3 & FP4 zones. In H2 - zone of fine silt and clay, chalky, similar to FP2 (see Lewis 1998 Appendix III.ii)
	H2	9-18	apedal crack, crumb, subangular blocky	15-20%: channels, cracks, vughs	25:75	stipple-speckled and crystallitic	10%: 'punctuations' and other fine fragments, seeds	none noted	

Table 23 Microscopic characteristics of tilled samples from Butser Ancient Farm: Spring-sown emmer for three years, non-manured and manured, spade-tilled and hoed, fork-turned in 3rd year

Profile	Subhorizon	Depth (cm)	Microstructure	Porosity (%)	c:f ratio	Groundmass	Organic components	Textural pedofeatures	Fabric pedofeatures
F3 1 **M** Ap	H1	0-7	subangular blocky, crumb, vermicular	20-30%: cracks, packing pores, vughs	20:80	stipple-speckled and crystallitic	10-20%: coarse degrading remains, spores, fine fragments, seeds, charcoal	all horizons: occasional silt/clay coatings and infillings	all horizons, especially H2 - FP3 zones (see Lewis 1998 Appendix
	H2	7-8.5	crumb, subangular blocky, laminar	20-30%: packing pores, cracks, vughs	25:75	as H1	15-20%: fine fragments, some carbonised, some coarse degrading remains		III.ii)
	H3	8.5-13	subangular blocky, crumb, vermicular	30-35%: packing pores, vughs	25:75	as H1	15-20%: as H2, also large root fragments		
	H4	13-18.5	crumb, subangular blocky, vermicular	as in H3	25:75	as H1	15%, as H2		H4 - FP2 zones (see Appendix III.ii)
F3 2 **M** Ap	H1	0-12.5	subangular blocky, crumb, vermicular	40%: packing pores, vughs, channels, cracks	25:75	stipple-speckled and crystallitic	15%: fine fragments, some coarser structural remains, some iron staining	occasional micritic clay coatings, earthworm sorted zones of silt/sand	FP3 & FP4 zones (see Lewis 1998 Appendix III.ii)
	H2	12.5-19.5	subangular blocky, crumb and granular, apedal crack	20%: cracks, vughs, channels	35:65	as H1	<10%: 'punctuations' and fine fragments	earthworm sorted zones of silt/sand	
F3 3 **NM** Ap	H1	0-11	subangular blocky, crumb, granular, vermicular	20-15%: packing pores, cracks, vughs, channels	20:80	stipple-speckled and crystallitic	10-20%: coarse degrading remains, spores, fine fragments, seeds, charcoal	none noted	H1-2 - FP3 & FP4 zones (see Lewis 1998 Appendix III.ii)
	H2	11-17	subangular blocky, crumb and granular, vermicular	15-20%: vughs, cracks, channels	30:70	as H1	10%: as H1, some very coarse cellular remains	linear infillings/coatings of fines	H2 - FP2 zones(see Lewis 1998 Appendix III.ii)
Ap/C	H3	17-19	irregular blocky, single grain	as in H2	30:70	as H1	<5%: 'punctuations'	none noted	none noted
F3 4	H1	0-4	subangular-angular blocky, crumb and granular, laminar	30%: cracks, channels, vughs	20:80	stipple-speckled and crystallitic	10-20%: coarse degrading remains, fine fragments, rare charcoal	none noted	all horizons: FP1, FP3 & FP4 zones (see Lewis 1998Appendix III.ii)
NM	H2	4-12	apedal crack, subangular blocky, crumb and granular	15-20%: vughs, channels, packing pores, cracks	20:80	as H1	as H1, more very coarse cellular fragments		
Ap	H3	12-25	crumb and granular	30%: packing pores, channels, vughs, cracks	15:85	as H1	10%: as H1, but very few coarse remains	18 cm depth - silt/clay pans	
C	H4	25-27	irregular blocky, crumb and granular, vughy	20%: as H3	25:75	Crystallitic	5-10%: fine fragments	none noted	

Butser Old Demonstration Area

Butser Old Demonstration Area is located at the base of a hill on calcareous colluvium, and was an experimental farm during the 1970-80s, with ard tillage since 1976 (Gebhardt 1990, 43). Reynolds (1987, 14, 16; 1989, 39, 55) describes the soil as being *c.* 30cm thick, comprised of rendzina mixed with clay-with-flints and chalk granules. It directly overlies Middle Chalk, and had an average pH of 8.2 (*ibid.*). Before the experimental farm was established, known arable activity was limited to some cereal production from the late 1940s to *c.* 1952. No chemicals were applied since 1972, and previous amendments were limited to nitrogen. After these land uses, the area was used for grazing up to the time of sampling in 1994 (Reynolds 1987; 1989).

Gebhardt (1990) included this experimental site in her research on the identification of ancient tillage using soil micromorphology, sampling four profiles in 1985. Her findings have been discussed briefly above regarding structure and textural and fabric pedofeatures, and manured *versus* non-manured arable plots. Additional information on environmental samples taken in 1990, before the site was abandoned by the Ancient Farm, is available in Allen *et al.* 1992.

Eight samples for soil micromorphology were taken in January 1994, for the purpose of examining remnant tillage features after several years of fallow; these represent three profiles from a field which was cultivated with an ard and hoe from 1976, with spring-sown wheat and a biannual manuring regime (Reynolds 1987, 17, 59; note, however, that this field was described by Gebhardt (1990, 44) as being manured every year). The soil was 30cm deep over chalk.

While all of the samples were taken from an old manured cultivated plot (Field VI; Table 24), when compared to Gebhardt's 1986 samples[2], they most resemble profile P4/section *1986/9* (cultivated, non-manured). Gebhardt's samples showed distinct structural differences between the uncultivated soil (1-2cm angular-subangular blocky, very dense peds), the cultivated and manured soil (large crumb and granular and subangular blocky peds), and the cultivated non-manured soil (small angular-subangular blocky and crumb and granular peds with high interpedal porosity). The latter was much more friable and broken up into small squarish peds. This is how the manured soil appeared after several years of abandonment, suggesting a decrease in organic matter in this manured soil since it was left for pasture.

All of the 1994 samples are totally different from the original untilled structure described by Gebhardt (1990). After lying fallow (under grass) for 4-5 years, the previously manured profile did not see a major increase in biotic activity, although slight variations in structure from Gebhardt's samples (greater porosity and smaller crumb, granular and blocky peds) might relate to some change in faunal activity or to season of sampling. There was no reversion to grassland soil in the period of time in question, although the samples do show evidence of grass rooting (fine subangular blocky structure).

Several possible microscopic tillage indicators remained in the Butser Old Demonstration Area samples after being abandoned to grazing (Lewis 1998). Given the experience of Butser Ancient Farm (above), it is unfortunate that no control samples were taken from this site. The abandoned arable profiles have to be compared to Gebhardt's profile P2 (untilled from 1986) (Gebhardt 1990).

Regarding tillage implement impact, there is practically no correspondence of features with the predicted depths of tool impact (Romans & Robertson 1983a; 1983b) (Tables 25 and 26), suggesting that even after only a few years of abandonment these zones may not be seen micromorphologically. A rooting zone was noted at 8-10 cm depth, but no clear association with tillage impact (such as with compaction or some other structural characteristic) was seen. There is, however, one exception: at 13cm depth in *AG3* there is a zone of clay features which resemble Bt horizon material, but are slightly 'dusty'. This was tentatively interpreted as a fabric pedofeature (Lewis 1998, app. III.iii), but it is possible that it is a textural pedofeature, given that Gebhardt's sample 1986/6 from the same plot includes two linear pedofeatures with a similar fabric in one zone. In Gebhardt's thin section these appear to be infillings, while in AG3 the fabric is more delimited as a zone (Lewis 1998). If these features are interpreted as being fabric pedofeatures, they could represent B horizon fragments. Fine fraction particle size analysis results (Lewis 1998, app. II) show the clay contents of the profiles sampled generally decrease with depth (except for profile *AG1*), and no B horizon was seen on the site during sampling. If this pedofeature is indeed textural, then there may be some significance to its accumulation at 13cm depth, which is immediately under the depth of ard tillage impact related to textural pedofeature accumulation by Romans and Robertson (1983a; 1983b). Richard Macphail's sample *1314* (1997, pers. comm.), taken from the site in 1990 shows frequent clay zones and infillings, which were not noted in the samples taken for this study.

Other features which correspond with those seen in Gebhardt's thin sections include mixing (probably related to faunal activity) and inclusions of chalk fragments in the uppermost horizons. No subrounded organo-mineral fragments (Gebhardt 1990, 45) were noted.

[2] I am grateful to Anne Gebhardt for permitting me to examine and discuss her thin sections here.

Table 24 Field descriptions of profiles from Butser Old Demonstration Area

Profile	Depth	Description	Samples
AG1	42cm	Ap horizon 35cm thick clayey silt with occasional small-medium angular and subangular flints, tiny rounded chalk fragments at base; B/C horizon 7cm thick clayey silt with occasional to moderate fine rooting, rare charcoal flecks; C horizon is crushed chalk	AG1/1-1/3
AG2	30cm	Ap horizon 30cm thick, as *AG1*; no B/C; C horizon is crushed chalk	AG2/1-2/2
AG3	35cm	As *AG2*	AG3/1-3/3

Table 25 Subhorizon boundaries and textural pedofeatures from Butser Old Demonstration Area samples, compared to predicted depth of impact of hoe and ard tillage (after Romans and Robertson 1983a; 1983b)

	Manured every 3 years			
Depth (cm)	**AG 1**	**AG 2**	**AG 3**	
0-2	Ap	Ap	Ap	
2-4				
4-6	Ap			Predicted depth of hoe tillage
6-8				
8-10	Rooting zone	Rooting zone		
10-12	Ap	Ap		
12-14			fab/text pf	Predicted depth of ard tillage
14-16				
16-18				
18-20				
20-22	Ap			
22-24				
24-26	stone line			
26-28	Ap		stone line	
28-30			A/C	

Little Butser

The first Butser Ancient Farm was located on a northern spur of Little Butser, *c.* 271m above sea level. The hill consists of Middle Chalk underlying clay-with-flints (Reynolds 1979, 19). The site has with a thin (*c.* 10cm) rendzina soil directly over chalk. Reynolds (1987, 6, 14; 1989, 35) describes the soil as having an average pH of 8 and as being prone to desiccation, becoming loose and powdery after an unseasonably dry summer. Excavations in 1973 suggested that there was no previous arable activity on the spur, and that the soil depth was constant for the past two millennia, based on lack of disturbance to the underlying chalk. Before the Ancient Farm was established, the area had been intensively grazed (Reynolds 1979, 19). Samples were taken for this study from four fields, corresponding with fields listed by Reynolds (1987); their history is summarised in Table 27.

None of the tilled samples possess features that could be solely attributed to tillage, but there are many features apparently related to tillage when compared to the uncultivated soil (FF1 8) (Tables 28 and 29). All samples had an unusual structure, consisting of widely-spaced, very well developed, 3-5mm angular to subangular blocky peds (Lewis 1998). The untilled sample does not have such widely-spaced peds as the tilled, but still has relatively wide-spread peds that are strongly accordant. Sampling took place at the end of a very dry summer, and a very friable structure appears to correspond to that described by Reynolds (1989, 35) as being typical for this location under dry circumstances.

The main difference between the tilled and untilled samples is that peds in the untilled soil have an internal crack structure which is strongly expressed and angular to subangular blocky, while the peds in the tilled soil have only a few cracks (Lewis 1998). The untilled sample is the only one to have clearly-expressed clay pedofeatures (grain coatings), although one zone of 'dusty' clay does exist in FF1 5. While it is possible that the untilled location may for some reason have better clay accumulation or retention properties, it is also possible that clay was lost in the tilled samples due to stronger leaching. A higher rate of leaching could be created through the combined impact of disaggregation and dispersion related to tillage and rainfall.

Profile	Subhorizon	Depth (cm)	Microstructure	Porosity (%)	c:f ratio	Groundmass	Organic components	Textural pedofeatures	Fabric pedofeatures
AG1 Ap	H1	0-3	mixed subangular blocky, crumb and granular, spongy	10-30%: cracks, channels, vughs	30:70	crystallitic	10%: fine fragments, occasional cellular and charcoal remains	none noted	all horizons: stipple-specked and chalky zones
	H2	3-10	mixed subangular blocky, crumb and granular	20-30%: as H1, but often larger voids	30:70		5-10%: as H1, some root pseudomorphs		
	H3	10-19	as H2	20%: channels, cracks, vughs	25:75		<10%: as H2, fewer fine fragments, more coarse cellular remains, including roots		
	H4	19-23	less well-developed subangular blocky, crumb and granular	20-30%: cracks, vughs, channels	25:75		<10%: fine fragments, some larger cellular remains		
	H5	23-28	as H4, becoming apedal vughy	as H4	30:70		<10%: fine fragments		
AG2 Ap	H1	0-10	mixed subangular-angular blocky, crumb and granular, some elongated peds	20%: cracks, channels, vughs	35:65	crystallitic	10%: fine fragments, some medium cellular remains	none noted	all horizons: zones as in AG1
	H2	10-22	mixed subangular blocky, crumb and granular	20-25%: as H1	30:70				
AG3 Ap	H1	0-27	mixed subangular-angular blocky, crumb and granular	10-30%: cracks, channels, vughs	30:70	crystallitic and stipple speckled	10-20%: fine fragments, some large cellular remains, rare charcoal	clay-rich aggregate, uncertain if fabric or textural pedofeature (see main text and appendix)	
A/C	H2	27-29	mixed crumb and granular, subangular blocky	30-40%: packing pores, channels, vughs, cracks	50:50	crystallitic	<10%: 'punctuations' and fine fragments, rare charcoal	none noted	zones as in AG1

Table 26 Summary of micromorpological descriptions from Butser Old Demonstration Area

The untilled sample has only a couple of relatively small chalk fragments, while those tilled with a hoe, spade and ard include a great number of angular chalk fragments, correlating well with Reynolds' report (1981, 99-100) that the ard ripped up the chalk at this site. This conforms to the idea that some untilled topsoils may be relatively stone-free, due to faunal sorting, while those experiencing mechanical disturbance and a relatively low level of faunal activity have numerous stone inclusions (e.g. Ashbee *et al.* 1979, 282). One exception to this is FF1 4, which, while spade- and hoe-tilled, does not have a clearly-defined horizon with larger chalk fragment inclusions. In fact, samples tilled with a spade and hoe only do not have as many large chalk fragments in general as do those tilled with an ard as well, perhaps reflecting the deeper interaction of the traction implement or simply an increase in disturbance level associated with an increased number of tilling operations.

The background sample has a relatively low birefringence and is stipple-speckled, while all of the tilled samples have a yellow (XPL) groundmass, which is crystallitic due to an abundance of micrite (Lewis 1998). While a higher degree of organo-mineral staining may mask micrite in the samples, the tilled profiles appear to have seen general relative calcification (*ibid.*). According to Hall (1990), erosion is the main mechanism for increasing calcium carbonate contents in tilled rendzinas.

Orange-yellow (PPL) amorphous organic remains are found in all of the tilled samples, both manured and non-manured, although, significantly, not in the lynchet or background samples (Lewis 1998). Possibly similar types of features may be related to amendment (J. Heathcote, 1997, pers. comm.; Gebhardt 1990, 45), although no correspondence with the experimental land-use regimes is evident. There are no remaining clear indicators of manuring (e.g. dung fragments), except for one fabric pedofeature at the base of FF1 5/1 (manured), which might be a degraded dung fragment (Lewis 1998). Goldberg and Macphail (2006, 265-6) note that some manured soils contained dung fragments in thin section samples taken in 1990. Whether this difference relates to some accident of sampling or represents destruction of dung fragments in the few years between the taking of these sample sets is unclear.

Compared to the other Butser sites, there are very few earthworm granules and vermiform features seen in the samples from these samples. This is surprising given the relatively calcitic nature of the soil, which is also not lacking in organic matter content compared to the other sites (Lewis 1998). It is unclear if this could be related to previous land use; perhaps this soil or location is simply not as well-suited to earthworms.

Little can be said regarding depth of implement impact zone as the profiles are shallower than the predicted depth for ard and spade disturbance, and the changes in the profile at the hoe tillage depth (Table 28) are mostly related to the base of the litter and upper rooting layer.

The lynchet sampled has a distinctive organisation, which includes downslope orientation and sorting of materials, including an oriented planar void with a silty clay infilling, and oriented stone lines. It is not as micritic as the tilled samples, although it is more so than the untilled profile. The upper horizons show strong organic staining, similar to that seen in the untilled sample. The lower horizons show amorphous sesquioxide staining. Many peds in the lower horizons have an internal fine crack structure that is similar to that in FF18 (above). In common with the tilled samples, however, the lynchet profile has no textural pedofeatures (besides the one oriented infilling mentioned above) (Lewis 1998).

Table 27 History of locations sampled from Little Butser

Field	**History** (after Reynolds pers. comm.; 1987, 15-8, 25, 36)	Samples
I	Spade, hoe & ard cultivation 1972-1987. Fallow until sampled in August 1994. Cereal/bean rotation, spring-sown crops (Celtic bean & emmer), no manuring.	FF1 2/1
II	Spade, hoe & ard cultivation 1972-1987. Fallow until sampled in August 1994. Bean/wheat rotation, autumn-sown wheat, no manuring.	FF1 3/1
III	Apparently the equivalent of Reynolds' Field V, which held flax crops in a 2-year rotation. Lynchet formation was noted. Two samples were taken, one from the side of the field towards the centre of the spur (FF1 6/1), from a profile 7-10cm deep, and the other from the lynchet on the edge of the spur (21° slope) (FF1 7/1-2).	FF1 6/1, 7/1-2
IV	Spade and hoe cultivation 1975-1987. Spring-sown wheat, manured (cow dung and straw) & non-manured regimes. Fallow until sampled in August 1994. 10cm profile depth, moderate to frequent chalk fragments, profile overlies large chalk blocks (10-15cm), mollusca noted in soil profile. Two samples were taken. According to field discussion, one was taken from manured and one from non-manured parts of the field, but comparison to Reynolds (1987, 36) suggests that both samples might come from manured strips. In the field, FF1 4/1 was thought to originate from a non-manured area. FF1 5/1 was known to be from a manured area; its profile is as 4/1, but 13-15 cm deep.	FF1 4/1, 5/1
n/a	Background sample. 20 cm profile depth	FF1 8/1

Table 28 Subhorizons from Little Butser compared to Romans and Robertson's (1983a & b) predicted depth of implement disturbance

	Untilled background	Spade-, hoe- and ard tilled			Spade- and hoe- tilled		Untilled, lynchet	
Depth (cm)	**FF1 8/1**	**FF1 2/1**	**FF1 3/1**	**FF1 6/1**	**FF1 4/1**	**FF1 5/1**	**FF1 7/1-2**	
0-2	L/Ah	L	L	L	L	L	L	
2-4	Ah	Ap	Ap	Ap	Ap	L/Ap	L/A	
4-6				Ap		Ap	A	depth of hoe tillage
6-8			Ap/C			Ap/C	stone line	
8-10							A	
10-12								
12-14								depth of ard and spade tillage

Lejre Historisk-Archæologisk Forsøgscenter

The Lejre Historical-Archaeological Experimental Centre, located southwest of Copenhagen (Denmark), has been described briefly in Chapter 4. Four fields (numbered 1-4) were previously subject to ard tilling experiments. It was possible to sample from one ard-tilled field (Field 3), and to obtain samples from Field 1 for comparison to the Donneruplund ard tilling conducted at Silsoe College (discussed in Chapter 4); see Lewis (1998) for field locations.

Field 1 is a long area running up and over a hill, believed to have been small fields (land use unknown) before the centre started. It was pasture for at least 30 years before sampling, except for a small experimental area at the base of the hill. This area was tilled with an ard once, and six plots laid out within it were tilled twice, all with a Døstrup ard replica. These areas were left fallow (untilled) since 1994 (M. Rasmussen, 1995, pers. comm.). Before sampling for this project it was discovered that the field had actually been broken using a mechanised plough (*ibid.*). The original intent of sampling from this field was to compare the 'untilled' soil with the once- and twice-tilled plots, and with the results of tillage experiments using a replica Donneruplund ard. A pair of oxen was trained in Field 1 during 1997-8 using a Donneruplund ard replica, and this training produced furrows which could be studied as comparisons to the furrows from the Silsoe experiments (see Chapter 4). The discovery that the field had been tilled with machinery means that the samples taken to look at profile indicators of tillage (*versus* ard furrow features) can be used only in comparison to each other, for differences between the areas once and twice ard-tilled, in comparison to the background modern tillage. The micromorphological features described cannot be taken to be indicative of ard tillage.

Field 1 was sampled in September 1995 and September 1996. The first year's sampling consisted of three profiles:

- LEJ1, located in a strip tilled twice with an ard and LEJ2, located in the area tilled once only. Both had a ground cover of high grass and thistles
- LEJ3, located on the crest of the hill, in an area which was always pasture up to the time of the first sampling.

In the second year, additional profiles were taken from this field, one from the twice-tilled strips (LEJ6), and one from the area tilled once (LEJ7). The twice-tilled strip had a ground cover of high grass and thistles, while that of LEJ7 was short grass. The change in cover compared to LEJ2 may relate to micro-topographic differences, as the slightly downhill position of LEJ2 had a high grass cover. The soil in this field is an argillic brown earth. One difference with the second set of samples was that samples were taken only down to near the base of the B horizon; the stony C horizon was not reached. Both profiles were exposed to 26cm depth (Lewis 1998).

The Field 1 samples represent typical grassland and tilled fallow soil profiles, but with some slight variation that may be related to the different histories of the areas sampled. The deeper A horizons and thinner B horizons seen in the ard-tilled profiles LEJ1-2, unlike LEJ3, are most likely related mainly to position on slope, and not to A horizon thickening under tillage, although the influence of tilling, downslope movement, rooting or burrowing on the profile was seen in the twice-tilled plot (LEJ1) through the presence of what appeared to be B horizon aggregates in the Ap (Fig. 14).

Table 29 Summary of micromorphological characteristics of samples from Little Butser

Profile	Subhorizon	Depth (cm)	Microstructure	Porosity	c:f ratio	Groundmass	Organic components	Textural pedofeatures	Fabric pedofeatures
FF1 2 *NM* L	H1	0-2	crumb and granular, subangular blocky	65%: packing pores, vughs, cracks	5:95	stipple-speckled and crystallitic	95%: coarse litter and root fragments, strong organic staining	none noted	none noted
Ap	H2	2-12	granular, crumb, subangular blocky	40-50%: cracks, channels, vughs	15:85	crystallitic	<20%: fine fragments, amorphous organics, coarse cellular remains		zone of non-micritic fabric
FF1 3 *NM* L	H1	0-1.5	crumb and granular, subangular blocky	60-70%: packing pores, vughs, cracks	5:95	stipple-speckled and crystallitic	99%: coarse litter, root fragments, fine fragments and amorphous organics	none noted	none noted
L-Ap	H2	1.5-2	crumb	40-60%: packing pores, channels, vughs, cracks	10:90	crystallitic	30-60%: as in H1		stipple-speckled and less micritic zones
Ap	H3	2-6	granular, subangular-angular blocky	40-50%: cracks, channels, vughs	15:85	crystallitic	20-30%: coarse and medium cellular remains, fine fragments, amorphous organics		
Ap/C	H4	6-12	subangular-angular blocky, crumb and granular	20%: cracks, channels, vughs	15:85	crystallitic	10-15%: as in H3, also rare charcoal		
FF1 4 **NM** L	H1	0-1.5	crumb and granular, subangular blocky	70-80%: packing pores, vughs, cracks	5:95	stipple-speckled and crystallitic	95%: coarse litter and root fragments, organic staining	none noted	none noted
Ap	H2	1.5-11	crumb, subangular blocky, granular	30-60%: packing pores, cracks, channels, vughs	20:80	crystallitic	15-20%: coarse and medium cellular remains, fine fragments, amorphous organics		zones of non-micritic fabric
FF1 5 ***M*** L	H1	0-1.5	granular, subangular blocky	60-70%: packing pores, vughs, cracks	5:95	stipple-speckled and crystallitic	95%: coarse litter and root fragments, strong organic staining	none noted	none noted
L-Ap	H2	1.5-2.5	granular, crumb, subangular blocky	50%: cracks, channels, vughs	10:90	crystallitic	50-60%: as above, also fine fragments and amorphous organics		rounded decalcified zones and peds
Ap	H3	2.5-6	granular, subangular blocky	40%: cracks, channels, vughs	15:85	crystallitic	10-15%: coarse and medium cellular remains, fine fragments, amorphous organics	zone of dusty clay lenses with micrite	
Ap/C	H4	6-9.5	granular, subangular blocky	25-30%: as in H3	15:85	crystallitic	as in H3	none noted	

Table 29 (continued) Summary of micromorphological characteristics of samples from Little Butser

Profile	Subhorizon	Depth (cm)	Microstructure	Porosity	c:f ratio	Groundmass	Organic components	Textural pedofeatures	Fabric pedofeatures
FF1 6 L	H1	0-0.4	granular, subangular blocky	60%: packing pores, vughs, cracks	5:95	stipple-speckled and crystallitic	85-90%: coarse litter and root fragments, strong organic staining	none noted	none noted
Ap	H2	0.4-2.5	subangular to angular blocky, crumb	50%: packing pores, channels	15:85	crystallitic	25-30%: fine fragments, amorphous organics, coarse cellular remains		
Ap	H3	2.5-12	subangular to angular blocky	40%: as in H2	20:80	crystallitic	10-15%: as in H2	stipple-speckled channel infill	very micritic zone
FF1 7 L	H1	0-1	granular	65%: packing pores, vughs, cracks	5:95	stipple-speckled	80-90%: coarse litter and root fragments, strong organic staining	none noted	none noted
L-A	H2	1-2.5	crumb, granular, subangular blocky	50-60%: packing pores, vughs, cracks	5:95	stipple-speckled	50-60: as above, also fine fragments and amorphous organics	lens of fine microaggregates	
A	H3	2.5-4.5	granular, irregular and subangular blocky	40-50%: cracks, channels, vughs	15:85	stipple-speckled and crystallitic	25-35%: fine fragments, amorphous organics, coarse cellular remains	none noted	zones of more crystallitic fabric
A	H4	4.5-14	subangular blocky, angular blocky, granular	30%: cracks, channels, vughs	15:85	stipple-speckled and crystallitic	<20%: fine fragments, amorphous organics		
FF1 8 L	H1	0-1	granular, subangular blocky	>60%: packing pores, vughs, cracks	5:95	stipple-speckled	95%: coarse litter and root fragments, strong organic staining	none noted	none noted
L/Ah	H2	1-1.5	crumb, subangular blocky, granular	50%: packing pores, cracks, channels, vughs	5:95	stipple-speckled	50%: fine fragments, amorphous organics, coarse cellular remains		
Ah	H3	1.5-12	subangular-angular blocky, granular	30-40%: cracks, packing pores, cracks, vughs	10:90	stipple-speckled	10-20%: coarse and medium cellular remains, fine fragments, amorphous organics	dusty, dirty and clean clay zones and coatings	

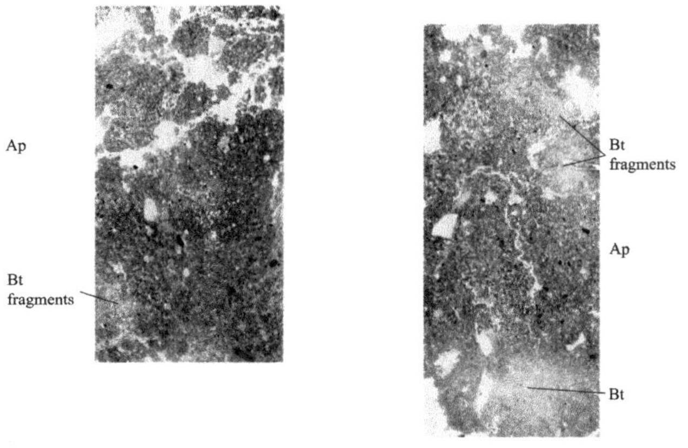

Figure 14 LEJ 4/1(left) and 4/2 (right) from Field 3 at Lejre; the light areas are yellow clods of Bt horizon materials, mixed into the Ap by tilling. LEJ 4/1 sample immediately overlies LEJ 4/2, representing the vertical soil profile as seen in the field trench (1:1scans; both thin sections are 5cm across)

Field 3 is in the Old Demonstration Area, which was tilled using a Døstrup ard replica for 10 years. No manuring was carried out, and the field was considered exhausted by 1994, when it was abandoned to fallow. Various crops had been grown on Field 3, but the last crop in the uphill area sampled was flax, some of which still provided ground cover when sampled in September 1995 (LEJ4) and in April 1996 (LEJ5). No further sampling was carried out, as the field was grassed over (Lewis 1998).

The soil in Field 3 was classified as an argillic brown earth in the field. It possessed a shallow Bt horizon and overlay chalk. Field 3 showed some characteristics of a tilled soil in the field; mixing was evident, with small patches of B-horizon material found in the topsoil (Lewis 1998). It is unclear whether these small aggregates represent tillage or harvesting - Pals and van Dierendock (1988, 242; after Dewilde 1983) list 'pulling' as the first stage in flax processing, and M. Rasmussen (1996, pers. comm.) suggested that the flax in Field 3 was harvested by pulling; these inclusions could have been pulled up with roots or through the action of the ard.

Field 1 (LEJ1-3, 6-7)

The depths of subhorizon boundaries and major pedofeatures are shown in Table 30. The identification of a 3cm deep ard mark unseen in the field in thin section LEJ6 (Fig. 9 in Chapter 4) suggests that, although some tilling may have been deeper, in this specific location at least depth of tillage can be compared with the results from the oxen training runs (Chapter 4). The known ard impact zone at Lejre under experimental tillage is thus roughly the same as the proposed depth of impact for hoe tillage (Romans & Robertson 1983a; 1983b). This variation, related to soil properties, climate and team efficiency (and possibly to implement), bodes poorly for any attempt to differentiate implement types based on depth of disturbance or subsequent textural pedofeature illuviation, except where previous land surface, disturbance factors, soil type and environment are extremely well-defined. In the other tilled samples from this field, boundaries and textural pedofeatures at this shallow depth are difficult to attribute to tillage as opposed to surface horizon influences. In sample LEJ7 a subhorizon boundary marked by a planar void or line of interconnecting vughs, and changes in structure from crumb, granular and subangular blocky to apedal vughy could, given the results from the experimental tillage at Silsoe, be interpreted as expressing a tillage boundary. Similar traits (without the textural pedofeatures) are seen at the upper subhorizon boundary in LEJ3, which was never tilled.

'Dusty' and 'dirty' clay coatings are found throughout the groundmass of all of the profiles, including the pasture soil, and as discrete concentrations at various depths in all except the untilled profile (Lewis 1998). The latter supports the association of these microfeatures with disturbance, including that created by tillage. Several discrete zones of 'dusty'-'dirty' clay coatings, including 'whole-soil' coatings (Fig. 15), were encountered. The latter appear at 15-25cm depth in LEJ2 and LEJ6, occurring slightly underneath Romans and Robertson's (1983a & b) predicted ard tillage impact zone only in LEJ6. They were seen both in the profiles that were broken with a mouldboard plough and in the profile that was only tilled with an ard and tractor. These might be compared to 'agric horizons', and may relate to mechanised tilling, although this was extremely short-term.

The profiles in general show an apedal (vughy) structure (Table 31); this is also seen in the pasture profile. There are some fabric pedofeatures (B horizon material in the Ap horizons) which may be related to tillage mixing of horizons, but these are also seen in the untilled profile. Despite being cultivated with modern machinery and an ard, the tilled profiles are only substantially different from the background samples in the discrete textural pedofeatures mentioned above. The minimal tilling has not apparently influenced structure (pedality or porosity), fabric, organic components or inclusions.

Field 3 (LEJ4-5)

This field was tilled for 10 years using an ard, and it shows much greater disturbance than Field 1. The relatively large fragments of B horizon material interpreted in the field as probably related to flax pulling (see above), may also be attributed to mixing associated with tillage in one of the samples (Lewis 1998). In the field they were seen to occur at 12cm depth, and are therefore some of the few features seen in the experimental sites which fit Romans and Robertson's (1983a; 1983b) predicted depth of ard disturbance (Tables 31 and 32). However, it is the lower part of subhorizon H3 (15-22cm depth) which shows the strongest mixing (Lewis 1998); this is deeper than would be expected using the ard (even in the Silsoe soil bin such depths were not achieved). Note also that this profile is from the crest of a hill, so it is not a naturally receiving environment.

'Dusty' and 'clean' clay pedofeatures are seen throughout the profiles, but there are two discrete areas of concentration. In LEJ4 there is a series of void coatings of very 'dirty' clay, silt and fine sand occurring at 5cm depth – this accumulation zone would be at the right depth for the ard disturbance seen in LEJ6 and 8-9, and it is similar in some ways to the 'pans'/linear infillings seen in LEJ8 and LEJ9 (see Chapter 4), with a relatively low birefringence and a very fine textured nature. In LEJ5, at 3-4cm depth, there are a great number of silt-clay coatings on peds, producing a marble-like fabric. This may be a 'whole-soil' coating (after Fitzpatrick 1993, 189); whether it relates to tilling ('agric') or to surface sealing is unclear (Fig. 15).

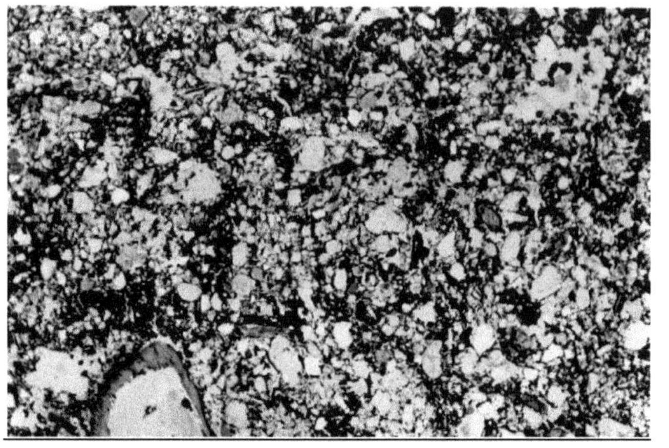

Figure 15 Possible 'whole-soil' coating in LEJ5/1 (PPL); arrow length represents 1mm

Table 30 Subhorizon boundaries and textural pedofeatures from Lejre Field I samples, compared to predicted depth of impact of ard tillage

Land use:	Ard tilled twice		Ard tilled once		**Pasture**	
	LEJ 1 – with modern plough	**LEJ 6 – with tractor and ard**	**LEJ 2 – mod. pl.**	**LEJ 7 – trac. ard**	**LEJ 3**	
Depth (cm)		Upper fill of				
0-2	text. pf	ard mark	A(p)	A(p)	Ah	
2-4	A(p)	lower fill		text pf	Ah	**ard impact**
4-6		A(p)		A(p)		**zone**
6-8						(as seen in LEJ
8-10	text pf		A(p)/B(t)	A(p)	Ah	profiles; see text)
10-12	A(p)					
12-14						**ard impact zone**
14-16		text pf			Ah/Bt	(after Romans
16-18		A(p)/B(t)	text pf			& Robertson
18-20			text pf			1983a ; 1983b;
20-22			text pf			Gebhardt 1990)
22-24						
24-26	A(p)/B(t)			B(t)		
26-28						
28-30						
30-32						
32-34						

KEY

text pf – at specific depths, discrete

subhorizon boundary (dashed line)

base of excavation (solid line)

Table 31 Summary of micromorphological characteristics from Lejre Fields 1 and 3 profiles

Profile	Subhorizon	Depth (cm)	Microstructure	Porosity (%)	c:f ratio	Groundmass	Organic components	Textural pedofeatures	Fabric pedofeatures
LEJ 1 A(p)	H1	0-7	apedal vughy, crumb and granular, subangular blocky, vermicular	20-30%: channels, vughs	50:50	stipple-speckled	20-50%: coarse cellular fragments, roots, fine fragments	dusty clay ped linings at 1.5-2 cm depth	none noted
	H2	7-23	apedal vughy, subangular blocky	15-20%: vughs, channels	55:45	stipple-speckled	10-30%: as in H1, with depth finer	rare slightly dusty clay coatings & infillings; some dirty clay coatings, especially at top of subhorizon	rare zones of grano-striated fabric
A(p)/B	H3	23-34	apedal vughy	10-30%: vughs, packing pores, channels	50:50	stipple-speckled, reticulate striated	5-10%: fine fragments, occasional structural remains	rare clay coatings and infillings – dusty and limpid; some silty clay zones	grano-striated clean clay zones
LEJ 2 A(p)	H1	0-8	apedal vughy, subangular blocky, crumb and granular, spongy	25-30%: packing pores, vughs, channels, planar void at boundary	45:55	stipple-speckled	5-10%: coarse remains with structure, fine fragments, rare charcoal	None noted	granostriated B horizon material inclusion
A(p)/B	H2	8-24	crumb and granular, subangular blocky, apedal vughy	20-30%: packing pores, channels, vughs	40:60	stipple-speckled, granostriated with depth	<10%: fine fragments, rare cellular fragments, charcoal, iron-replaced remains	fine silt-clay coatings – possible 'whole-soil' coating especially at 16-24 cm depth; dusty-dirty infillings; clean laminated infillings and coatings	granostriated B horizon material inclusion
LEJ 3 Ah	H1	0-1.5	subangular blocky, vermicular	20-25%: packing pores, vughs, channels	50:50	stipple-speckled	20-30%: coarse cellular fragments, roots, fine fragments	silt and sand lensing and infillings	none noted
	H2	1.5-8	crumb, apedal, vermicular	20-25%: as H1	45:55	stipple-speckled	15%: fine fragments	silt and sand infillings (vermicular); infillings and coatings of dusty/silty clay, some clean infillings	granostriated B horizon material
	H3	8-14	crumb and granular, apedal vughy, subangular blocky	20-40%: packing pores, channels, vughs	45:55	stipple-specked, granostriated	<10%: 'punctuations' and other fine fragments, rare charcoal		
BT	H4	14-21	as H3	15%: as in H3	40:60	stipple-speckled, granostriated	as H3		

Table 31 (continued) Summary of micromorphological characteristics from Lejre Fields 1and 3 profiles

Profile	Subhorizon	Depth (cm)	Microstructure	Porosity	c:f ratio	Groundmass	Organic components	Textural pedofeatures	Fabric pedofeatures
LEJ 4 Ap	H1	0-5	subangular to angular blocky, crumb and granular	20-30%: channels, vughs, planar voids	45:55	stipple-speckled	15-20%: coarse structural remains, fine fragments	none noted	fragments of Bt horizon material, inclusions of calcitic (chalky) material in H2
	H2	5-9	apedal vughy, subangular blocky	10-20%: vughs, channels	45:55	stipple-speckled	20%: as above, rare charcoal	clean clay coatings, lenses of dirty clay, including one possible pan on top of horizon	
	H3	9-22	apedal vughy, apedal channel, angular blocky	20-30%: vughs, channels, packing pores	35:65	stipple-speckled, granostriated	5-15%: occasional coarse, oxidised remains, mostly fine fragments	granostriated clay zones	Bt material fragments
B/C	H4	22-24	apedal channel, apedal vughy	5-15%: vughs	25:75	crystallitic	<10%: fine fragments		
LEJ 5 L	H1	0-0.5	Crumb and granular	70%: packing pores, vughs	25:75	stipple-speckled	50-60%: iron stained coarse remains with cell structure, fine fragments	none noted	none noted
Ap	H2	0.5-12	apedal vughy, crumb and granular, subangular blocky, vermicular	10-20%: vughs, channels	20:80	stipple-speckled	10-15%: mostly fine fragments, some coarse as in H1; root running across section	at 3-4 cm depth – silt coatings – marbled appearance – agric horizon? sealing?	mixed chalky and clean clay zones (H3 fabric)
Ap/(B(t))	H3	12-23	apedal vughy, apedal channel	10-20%: vughs, channels	30:70	stipple-speckled	as H2	slightly dusty clay infillings and coatings	none noted
Ap/B(t)	H4	23-30	apedal vughy, apedal channel, zones of subangular blocky	15-20%: vughs, channels	45;55, 30:70 with depth	stipple-speckled, granostriated, porostriated	<10%: 'punctuations' and fine fragments		

Table 31 (continued) Summary of micromorphological characteristics from Lejre Fields 1and 3 profiles

Profile	Subhorizon	Depth (cm)	Microstructure	Porosity	c:f ratio	Groundmass	Organic components	Textural pedofeatures	Fabric pedofeatures
LEJ 6 upper fill	H1	0-1.5	crumb and granular, subangular blocky	20-40%: packing pores, cracks, vughs	45:55	stipple-speckled	10-20%: 'punctuations' and other fine particles, some structural remains	none noted	none noted
lower fill	H2	1.5-3	subrounded aggregates	40%: packing pores, vughs	35:65	as H1	as H1, but also large root remains		
Ap	H3	3-15	angular-subangular blocky at top, subangular blocky, crumb and granular, vermicular	10% under ard mark cut: vughs and cracks; 20% rest: vughs, channels, cracks	25:75	as H1	15:20%: as H2, but frequent medium-sized remains; horizontally-oriented remains under ard mark cut	lenses of fines coating voids, becoming complete infillings at 15 cm depth – akin to agricutans?	
B(t)	H4	15-28	apedal vughy, apedal channel, subangular-angular blocky	<10-35%: vughs, cracks, packing pores	40:60	stipple-speckled and granostriated	as H3 main horizon	infillings and coatings of clean and slightly dusty clay, some iron-stained	
LEJ 7 Ap	H1	0-2	crumb and granular, subangular blocky	20-25%: vughs, channels, cracks	40:60	stipple-speckled	10%: medium-coarse structural remains, fine fragments	in all horizons: very dusty clay and silt infillings and lenses	none noted
	H2	2-7	apedal vughy	10-15%: vughs, channels	30:70	as H1	10%: fine fragments, some structural remains	H2 – dusty-very dusty clay at top of horizon, possibly a weakly-developed 'agric' zone, increasing in expression with depth	none noted
	H3	7-23	apedal vughy, subangular blocky	15-20%: vughs, channels, cracks	30:70	as H1	as H3	H3 – rare clean to slightly dusty clay zones	subrounded (excremental?) aggregates
B(t)	H4	23-27	apedal vughy, apedal channel	10-15%: vughs, channels	30:70	stipple-speckled, grano- and porostriated	10%: fine fragments		

Table 32 Summary of textural pedofeatures and subhorizon boundaries over depth in Lejre Field 3 samples (LEJ 4-5), compared to predicted depth of ard impact

Depth (cm)	Profile: LEJ 4	Profile: LEJ 5	
0-2	Ap	L	
2-4		text pf	ard impact zone (as seen in LEJ profiles – see text)
4-6	text pf/pan	Ap	
6-8	Ap		
8-10			
10-12	Ap		
12-14		Ap/(B(t))	ard impact zone (after Romans and Robertson 1983a & b; Gebhardt 1990)
14-16			
16-18		15-22 cm strong mixing	
18-20			
20-22			
22-24	B/C		
24-26		Ap/B(t)	
26-28			
28-30			

Summary

A review of indicators of depth of implement impact and of characteristics of experimentally tilled soils suggests that the profile and horizon indicators summarised in Chapters 2 and 3 above do represent the impact of tillage, but that this is often difficult to relate to tillage itself, *versus* other processes, and that predicted zones of impact do not match actual zones of impact seen in many of the soils. When horizon or subhorizon boundaries do match those predicted by previous studies, it is possible that the defining characteristics (usually structural) relate to implement impact depth. However, there are many other aspects that could be involved; in particular regarding structural alterations related to rooting and faunal bioturbation, the relationship between tillage and the preferences of specific biota is difficult to untangle. For instance, these boundaries are probably strongly related to soil chemistry, which may be strongly affected by tillage events over both the short and long-term (archaeologically), but may also depend on many other influences, *i.e.* the relationship between biological activity and ancient (or even modern) cultivation is a very complex one that is not easily untangled. These indicators, as found by Gebhardt (1990) amongst others, are obviously relevant, but not in a clear-cut manner. Adding long time scales and changing environments to the picture makes these indicators even more complicated, although in certain environments these may be more clearly related to ancient tillage than in the biologically-active brown earths studied here.

The control samples from Butser Ancient Farm show features coinciding with the predicted depths of spade and hoe tillage, which makes it difficult to adequately assess claims that depth of impact can be determined and used to differentiate between these digging implements (e.g. Romans & Robertson 1983a; 1983b), and suggests that there may be a great deal of 'noise' in this regard in both modern and ancient profiles. Some of this background disturbance appears to relate to rooting and faunal activity zones. As noted in previous chapters, these biologically-created zones may also relate to depth of tillage, but they are not necessarily primary indicators of implement impact.

The degree of faunal reworking on samples from different seasons at Butser Ancient Farm is strong, and suggests that many characteristics caused by tillage may be unrecognisable within as little as one year of biological activity from the date of tillage, even in a frequently tilled soil. This was a predictable result from an earthworm-rich rendzina soil, and has some important implications for the soil micromorphological study of tillage in buried soils of this type. It may not be possible to positively identify ancient tillage from soil structure in (previously) biologically active rendzina soils. It may also not be possible to eliminate tillage processes in buried rendzina soil horizons which appear to be 'grassland/pasture' on the basis of structure. Many of the tilled samples from Butser show a structure dominated by faunal processes, but they were tilled, and not only with low impact implements. The continuum of increased faunal influence over the seasons mentioned previously suggests that major alterations in structure occur within one year (or even one summer) from last tillage in these soils. Where buried soil horizons are examined in similar environments (e.g. under monuments on chalk; French *et al.* 2007), it would thus seem that the only clear means of determining that a horizon represents a grassland soil that has never been tilled is to find evidence of turf. Even where such evidence is found, it would be foolhardy to conclude that no previous tillage occurred before turf formation on the basis of structure or the absence of textural pedofeatures such as infillings or even crusts. Regarding the latter, although they are most frequently found in tilled profiles, they are not restricted to tilled or even bare fields in this environment, and should not be used on their own to infer tillage or clearance from limited numbers of prehistoric samples from similar soils.

Despite all of this negative evidence regarding the usefulness of soil micromorphology in characterising or identifying ancient tillage on brown earths over chalk, it has been possible to distinguish between infilling/'panning' and crusting/sealing features micromorphologically at Butser Ancient Farm, and to suggest general characteristics of silt:clay pans that appear to relate to implement impact zones at two experimental sites. At Butser Ancient Farm, textural pedofeatures with characteristics that appear to be distinctive of 'panning'/infilling zones were located in these highly reworked soils, and it was possible to distinguish these from crusts. Whether these features are related to modern or experimental tillage was impossible to discern. Nevertheless, their existence in such soils and the fact that they were located micromorphologically is encouraging, although the chances of them surviving over the long term (*i.e.* archaeologically) are perhaps low. With intensive sampling regimes such as that carried out at Butser, however, it may be possible to find tillage-related textural pedofeatures even in highly reworked environments (c.f. French 2003; Lewis 2002). It is important to note that with a less-intensive sampling regime, the chances of finding these features would be very slim in soils such as those at Butser.

The lack of distinctive structures or fabrics related to manuring is important. Manures used at Butser Ancient Farm did not result in micromorphologically visible increases in organic matter content or faunal activity. The only inclusions which could be directly associated with manuring (dung fragments) were found in a non-manured plot. This result was surprising, as it does not fit the majority of experimental evidence (with modern or reconstructed ancient cultivation regimes, e.g. Goldberg and Macphail 265; see Chapter 2 above). It is not clear why manuring showed no visible impact on the soils at Butser studied here; other analytical approaches not employed in this study may have recognised this difference more clearly. In a previous study of the experiments at Butser, microscopic dung fragments were found in the manured fields, and associated with enhanced phosphate results (Macphail *et al.* 2000; Goldberg and Macphail 2006, 265).

At the Old Demonstration Area it was possible to compare a number of thin sections taken at different sampling times from the tilled site. This showed that several possible microscopic indicators of tillage identified in 1980s by Gebhardt (1990) continued to survive ten years later, although it was difficult to directly relate most of them to the specific tillage regime.

The results from Little Butser demonstrate several points. This site is one which shows textural pedofeatures often related to tillage ('dusty' clay coatings) in an untilled profile, while the tilled samples do not possess such pedofeatures. Tillage is distinguished here in comparison to the untilled samples by a lack of these features, a less well-developed ped structure, the presence of many large chalk fragments in the profile, a crystallitic as opposed to stipple-speckled fabric, the presence of amorphous organic matter, and a lower degree of organic staining. The amorphous organic matter content does not appear to distinguish manuring from non-manuring (being found in both types of profiles), but does characterise the tilled *versus* the non-tilled profiles. The site shows a number of classic characteristics of tilled soils, as discussed in Chapter 2, but not others; 'dusty' clay, for example, is only seen in the untilled sample. This lack could be explained through decalcification and/or simply extreme leaching without a suitable situation for illuviation; it provides yet another example of how important it is to appreciate specific soil properties in particular locations, instead of applying broad-brush models when interpreting land use from proxy indicators.

The samples from Lejre suggest that while Romans and Robertson (1983a; 1983b) are probably correct in assuming that textural pedofeatures will accumulate at the base of tillage, what that actual depth is may vary greatly even when only one implement type is involved. Textural and structural pedofeatures possibly associated with the depth of ard disturbance at this site occur, but they are located within the predicted depth of hoe tillage according to Romans and Robertson. This suggests that while these indicators are useful in interpreting where in the profile a base of cultivation depth may occur, they cannot be easily related to tool type in the absence of other indicators (e.g. marks, 'turning' indicators (Lewis 2002)).

The Lejre field tilled only with an ard (Field 3) shows more features which are clearly characteristic of tillage than does that tilled with both modern equipment and an ard (Field 1). The difference may be a factor of original soil profile and catenary position, and/or variation in soil turnover rate. Field 3 has a well-developed Bt horizon that was intersected by the tilling zone, *i.e.* the topsoil was relatively shallow, probably due to its location on the crest of a hill. This interaction with the B horizon is responsible for the creation of many fabric pedofeatures of mixing. The relative shallowness of the topsoil, along with any chemical changes associated with the bringing up of Bt and chalky material may have had an influence on the level of biological activity, *vs.* Field 1, which shows strong faunal disturbance of many types (Lewis 1998). The level of bioturbation in Field 1 may have limited the preservation of tillage indicators. The ten-year duration of the ard tillage in Field 3 may also have played a role in the stronger expression of tillage indicators. Regarding textural features, although in these samples (especially in those from Field 3) many 'dusty'-'dirty' clay pedofeatures were seen, it proved difficult to relate them to 'predicted' implement impact zones, although they do correspond with actual depth of tillage seen on the site. Despite the relatively high amount of clay in this soil, and clear indicators that translocated material does accumulate within the horizons examined, most translocated clay in pores or on grains was either clean or only slightly 'dusty', with 'dirtier' clay being found mainly in the groundmass (Lewis 1998).

The results presented in this chapter suggest that Romans and Robertson's (1983a; 1983b) ideas on accumulation zones can be supported in principle, but are extremely problematic regarding practical application. The difficulties discussed are seen even on sites where implement(s) of tillage and arable system are known, and in cases where there is little to no significant post-depositional alteration or profile truncation relative to archaeological situations. According to these results, it is unlikely that implement of tillage can be reliably or clearly related to specific depths in the soil profile based on either pedofeature occurrence or structural alterations in the absence of implement marks. This is the case despite the fact that such characteristics are expressed at the base of tillage marks (see Chapter 4).

Nevertheless, certain features (e.g. silt-clay 'pans'/infillings) were located in highly biologically active soils through intensive sampling, and certain combinations of field and microscopic profile characteristics could be seen to reflect differences in tilled and untilled profiles. These included:

- mixing indicators (e.g. rock fragments and B horizon materials pulled up into Ap profiles, and a lack of these in comparable Ah profiles),
- groundmass characteristics (e.g. organic stained and/or less calcareous fabrics in untilled profiles *versus* micritic and/or less organic fabrics in ard-tilled profiles)
- amorphous organic inclusions (in tilled profiles, only at Little Butser)
- general textural pedofeatures (e.g. at Lejre, where a good deal of clay is present in the soil, possible 'agric' features were seen in the tilled profiles; also the 'panning'/infilling features mentioned above, although these are held to relate more to implement mark characteristics. However, at Little Butser 'dusty' clay pedofeatures were only encountered in the untilled samples; while this was a distinguishing factor between tilled and untilled profiles, and is explicable through local soil characteristics, it is the opposite of what is normally expected).

Indicators which did not reflect cultivated profiles clearly micromorphologically include manuring indicators and earthworm activity. In addition, crusting, which is normally associated with bare earth processes, was seen in one untilled grass-covered profile. Structure, while reflecting tilling, was not a particularly clear indicator, perhaps because of the relatively high level of biological activity in these soils. At Butser Ancient Farm it was clear that a tilled structure can be very quickly reworked into a more faunally-dominated structure, throwing doubt on the possibility of potentially identifying both buried tilled and untilled (grassland) rendzina soils on the basis of structural characteristics, even where intact turf remains are also present. These results are both positive and negative in terms of the potential for using soil micromorphology as a method to identify and characterise ancient cultivation.

Chapter 6 Relating experiments to archaeology

Introduction

The characterisation of tillage features and implement impact zones is an informative means of applying the soil micromorphological approach, providing a way of investigating general and implement-specific tillage feature characteristics, and exploring the creation and location in the soil profile of certain profile and horizon characteristics associated with cultivation using ancient implements.

The most distinctive micromorphological features of tilling seen in the experimental samples discussed above were those associated with implement feature cuts. These zones mark structural and textural boundaries within the soil profile, and may retain their distinguishing characteristics over long periods of time. The cuts of the tillage features examined were always marked by at least two characteristics: aligned pore space along the cut itself (planar voids, shear planes, lines of packing pores), and a density change between the 'inside' and 'outside' of the cut (loosening inside, compaction outside). Other characteristics include associated distortion under the base and sides of the cut, and an accumulation zone for infillings of fine particles and small aggregates at the base of the cut itself (internally to or marking the cut) or immediately outside of the cut. These features were seen to be created both by direct implement action and through secondary infilling of void spaces. In some examples, rounded or flattened aggregates were also seen at the implement mark cut.

This chapter focuses on what impact these characteristics might have in the long-term. Experimental reconstructions of ancient tool impact in a modern setting are very useful for understanding how an implement functions in a physical context at least potentially similar to that for which is was designed. In particular it is an invaluable experience to see and feel an implement move through the soil, observing how structure and organisation are altered, and then to compare the soil micromorphology created with that documented and proposed for modern and ancient counterparts. Whether or not such experiments come close to past practices is, while not irrelevant, not of primary concern regarding issues of evidential interpretation of processes at a local scale. Taking such evidence beyond this level, into the realms of ancient social or palaeoenvironmental interpretation, is another matter entirely, and one which has been discussed elsewhere (e.g. Lewis 2010; 2008; French *et al.* 2007; Carter and Davidson 1998; Macphail 1998a; see also Coles 2010). Continuing in this volume with a feature-specific approach, there are still issues of taphonomy and representation to consider, and these are the subject of this chapter.

Structural zones created by implements

In the soil bin tilling experiments it was possible to show break up of soil into two basic types of aggregates: angular blocky clods of a relatively large size, mostly found deposited in ridges and occasionally in ard mark fills, and small, more rounded aggregates frequently found in the ridges and comprising much of the loose of the furrow. These types and arrangements of aggregates sound very similar to those resulting from modern tillage using vertical and forward raked narrow tines, which bring larger clods to the surface and leave smaller ones below (Spoor 1969b, 18), and to the findings of Gebhardt (1990, 44-6, 63-6) in experiments with ard tillage (see also Dexter 1976; 1979; Macphail *et al.* 1990a). In situations where untilled comparative material is available, such patterns might be recognised archaeologically. However, variations in soil type and subsequent land use will add an element of uncertainty to interpretations of ancient tillage action based on this aspect of soil structure. The results of the study of experimental fields show that cultivated structure is quickly altered in biologically active soils like rendzinas, and literature review suggests that it is difficult to easily interpret ancient land use on the basis of structural indicators, although these are clearly important characteristics of tilled zones in soil profiles.

Mixing of surface and subsurface horizons has frequently been cited as a micromorphological indicator of ancient tillage (e.g. Macphail 1992b; Currie 1994; Lewis 1998; Lewis and French 2006). The experiments presented above showed that aggregates from the lower part of the implement impact zone are pulled up into the surface of the furrow, and that fragments from upper layers drop down into the furrow fill. Soil from greater depths was seen to ride up the main share of the ard at Silsoe, mixing with materials from all depths both on the share and upon deposition. Furrow fills tended to show larger aggregates in the uppermost zone, followed underneath by a zone dominated by fine aggregates and disaggregated apedal material. These variations in structure and related porosity (density) appear to play a role in the deposition of material later translocated through the profile. Lateral and sideways movement of aggregates was also noted, with soil tilled by the ard moving ahead of the implement, following the notion of a soil failure zone or disturbance crescent (Payne and Tanner 1959, 312, 316-8), and in an upward and sideways displacement pattern. Ridges were seen to echo furrow fills in their structural organisation, but with zones following the diagonal direction of the ridge shape. Whether any of these arrangements will survive in archaeological situations depends on soil types, and burial and preservation environments. It is clear from archaeological examples mentioned throughout this volume that when a different horizon or layer was interacted with by an implement,

mixing of clods can be seen both macroscopically and in thin section.
Perhaps the most important structural alteration for archaeology is the creation of compaction or compression zones under and around implement disturbance zones (marks or features). Immediately underlying the mixed loosened zone of fills, basal zones of ard marks studied showed evidence of physical compaction and/or compression related to implement passage. In most cases, porosity of the uppermost part of the underlying layer was reduced, both in comparison to furrow fills and to immediately underlying material. In addition, a line of >1cm-sized 'rolled' or flattened aggregates was seen to underlie the furrow base and sides, particularly where an arrow-shaped share was used. This feature further enhances the impression of a structural boundary created between fill and cut. The location of this compaction zone within the first centimetre underlying the furrow is in line with findings based on modern tillage (Jongerius 1983, 122).

Shear planes were also observed to be created around implement mark cuts, both running off at right angles to the sides and echoing the cut under the base of the marks. It is suggested that this represents the impact of downwards and sideways pressure under the action of the implement. Shear planes created around the cut are very important to the location of microscopic features related to secondary processes such as illuviation, and will be discussed further below.

The opinion that subsurface compaction is not created by implements such as ards, except possibly as an effect of associated animal trampling (e.g. Carter and Davidson 1998, 542) is not supported by these results. The basic physics involved should lead us to expect to encounter compacted and compressed zones around all dug features – indeed, this is what we see when excavating features in the field, although it is not always the most obvious of the various characteristics involved in the creation and identification of a feature 'cut'. The impact of trampling has not been discussed in depth in this volume because it was beyond the scope of the research project; the compaction zones described here were created in situations with no trampling (in the soil bin) and were clearly related to the implement itself.

Finally in regard to structure, 'rolled' and flattened aggregates located at feature cuts were only seen in studies of traction implements; pending future study, these aggregates are assumed to relate to compaction in the zone where soil fragments are run over and rolled along by the base of the forward-moving implement.

Textural 'panning' and infilling pedofeatures in relation to structural zones created by implements

Through the soil micromorphological study of ard marks it has been possible to show that structural changes created by compaction and shearing can be related to specific types of textural pedofeatures at and around feature cuts. Comparison to the proposed accumulation depths for textural pedofeatures under different implements (Romans & Robertson 1983a; 1983b), supports the idea that translocated fine fraction materials accumulate at the base of a tillage disturbance zone. Indeed, in experiments at Lejre and from archaeological examples (e.g. Lewis 2002), illuviation has been seen to occur both at the bases and sides of implement mark cuts, and the most distinctive pedofeatures (infillings/'pans') are those directly associated with feature cuts (within a 1cm zone on either side). This, it has been suggested, results from and augments the density boundary (textural and structural) already created by the passing of the implement through the soil (Lewis 1998). These results concur with expectations from modern tillage practices, suggesting that we are correct in assuming that ancient tillage has similar, if possibly less strongly-expressed, immediate influences on the soil.

The fact that a lower layer of fine particles and aggregates was created at the bases of ard mark fills at the Silsoe Soil Bin, in soil which was not exposed to the elements and saw no in-profile translocation, shows that zones of fine materials are produced in this part of tillage features by the action of the implement itself, *i.e.* that they do not require post-depositional forces to exist. They are created by the physical impact of the implement, plus gravity, and are thus primary tillage feature characteristics. However, as the experimental ard marks from Lejre show, there is also a substantial and relatively quick-acting illuvial element involved in the creation of basal lenses of fine soil particles in features created in the field, at least in suitable soils. The creation of a textural and density boundary at this point in the profile should be expected to then play a major role in later secondary characteristics, such as the development of iron and clay pans, and possibly, in less bioturbated soils, even in soil horizon creation in the field. This supports Romans and Robertson's (1983a & b) fundamental suggestion that zones of accumulation of fine particles seen archaeologically might be interpreted as the base of ancient tillage, at least if they accumulated in the manner of lenses of fine fraction within void spaces created by implement impact.

Archaeologically, iron panning and silt-clay infillings have been described at the base of tillage layers at several sites, including the comparatives mentioned in previous chapters. One further example that may serve to demonstrate these relationships is that of Bjerre Site 7 in Jutland, Denmark (*c.* 57°3'0" N, 8°42'0"E; Bech 1993; Aaby 1993; Andersen 1993; Clemmensen *et al.* 2001), where certain horizon boundaries follow ard mark cut bases and sides in the podzolising soils of this site.

At Bjerre various layers of sand deposition and soil development were seen, including a series of layers of ard marks and cultivation soils dating from the early Bronze Age, and late Bronze Age field systems consisting of small, irregular plots (Bech 1993). The site lies on the

Littorina sea bed (*ibid.*), with acidic sands and repeated formation of podzolised soils interrupted by periods of erosion/aggradation and tilling. In the upper metre of the site profiles the classification of ferro-podzol is appropriate (Lewis 1998).

Thin section samples from the site consist almost entirely of fine and medium sand, with layers and horizons varying mainly in their amorphous organic matter and sesquioxide contents (Tables 33 and 34), representing both the development of topsoil horizons during phases of stabilization, and podzolisation processes (Lewis 1998). Several buried tilled horizons were identified across the site; all of these buried Ap horizons were distinguished in the field by their colour (grey, greyish brown or greyish yellow), and by inclusions of charcoal, pottery and flint artefacts. Several had associated underlying ard mark layers (*ibid.*). These ancient Ap horizons fit well with Liversage and Robinson's (1992-3, 45, 51) descriptions of late prehistoric buried soils associated with ard marks at other sites in north Jutland.

The ard marks at the site were generally about 6-10cm wide, with shallow and pointed, or deeper and rounded cuts in profile; some had an uneven profile, possibly created using a tilted ard. Most were very shallow (<2cm), but several had associated possible ridges 10-12cm wide and of variable height (Lewis 1998). In one location ard marks cut into an underlying 'culture layer', thereby tilling it as part of a later Ap horizon; some criss-cross tilling was noted in this set of features (*ibid.*).

Several thin sections were studied of all of the buried soils and of individual tillage features from the site (Lewis 1998). In thin section it was possible to identify additional, lower, ard marks unseen in the 'culture layer', suggesting that several episodes of tilling had occurred before the set of ard marks seen in the field; the latter were identified because they were infilled with blown sand, and stood out clearly in the black 'culture layer'. Deeper ard marks were characterised as V-shaped, oriented infilled planar voids, filled with translocated, dense black and brown amorphous organic matter with clay, invisible to the naked eye (Lewis 1998). As discussed in Lewis (2002), certain layers of implement marks appear to stand out in the field due to either their role in soil horizon boundary creation, to the truncation of a soil horizon, and/or to their being the last tilling event before burial (e.g. under blown sand or archaeological monuments). Detailed examination shows, however, that these apparently single event horizons can usually be shown to be part of ongoing cultivation processes, and that taphonomic explanations can easily explain the preferential survival of individual layers of marks (Lewis 2002; 1998); where examined in detail they are not usually related to 'ritual ploughing' or ground-breaking events, but more frequently linked to profile truncation (e.g. French *et al.* 2007) or sudden burial; demonstrating that a particular set of ancient cultivation marks represents a single tillage event requires a great deal of investigative focus on soil history and the impact of later truncation on feature visibility.

At Bjerre one layer of ard marks identified in the field was shown to be part of horizon creation processes involved in the podzolisation of the site. A layer of amorphous organo-mineral coatings related to (later) podzolisation was found in thin section as monomorphic coatings on grains running along the lower edges and up the sides of implement mark cuts (Figs. 16 and 17; Lewis 1998). This forms a small placic horizon; similar layers were seen in the uppermost part of horizons underlying cultivation soils, even creating iron pans visible in the field at one such boundary (although not associated in that location with specific implement marks).

Figure 16 Ard mark with leached (light) fill and cut demarcated by amorphous organo-mineral panning, mainly of amorphous iron. In colour, the fill is grey and the cut edge is reddish brown (see Lewis 1998). Thin section size 6x10cm.

Iron panning features require there to be strong leaching, occurring in soils seeing podzolisation processes; these soils tend to be sandy, acidic and relatively freely-draining with relatively low biological activity. None of the experiments described in this study occurred on such soils, and it is unclear how long it takes for such soil features to develop. As far as the author is aware, there has been no discussion of the location of placic characteristics marking the actual cuts of tillage features, instead of simply the base of a tilled horizon, but the logic behind the accumulation zone is the same as described above: loose implement mark fills next to a compacted surrounding soil matrix create a boundary zone at which such accumulation might be expected to occur (Fig. 17). Reynolds and Catt (1987, 515) suggest that the panning horizon seen at Dark Hat formed 'through compaction when the soil was cultivated at a moisture content close to or above field capacity', decreasing the permeability at the base of tillage. At Bjerre it seems likely that the pans formed well after the ard tillage episode involved (Lewis 1998).

In soils with more subtle characteristics than the podzolising sands at Bejre, such as those comprising the bulk of the experimental study reported in this volume, other types of pedofeatures are seen to mark the 'cut' of tillage implement features. One group of textural pedofeatures was seen here to be specific to 'panning' characteristics at the base and sides of implement marks in brown earths. These are silt/clay and very fine to fine sand 'pans'/infillings, as described above from furrow bases at Lejre and seen in certain profiles at Butser Ancient Farm. As Gebhardt noted in her experiments with an ard at Hambacher Forst (1990), organic-stained silt-dominated textural pedofeatures (60:40 silt:clay) were created in linear void spaces: in both cases these had a clear morphology related to tillage with an ard, and appear to be created relatively quickly in some cases. The main soil micromorphological characteristics of these discrete panning features are:

- A composition of mixed very fine microaggregates (<50-100μm, rounded or subangular blocky) with fine and very fine sand, silt grains and clay. The microaggregates seen appear to either relate to physical movement of these to this location under implement action (as at the Silsoe soil bin) and/or post-depositional aggregate formation or cracking within the 'pan'/infilling. In the case of podzols, such as those at Bjerre Site 7, the presence of microaggregates in pans was seen to relate to the creation of polymorphic coatings (Lewis 1998).
- A texture that appears to contain more fine silt than clay (*c.* 60:40), although the pans are usually organic- and/or sesquioxide-stained, making clay proportion difficult to estimate visually
- A marked lack of sorting into micro-layers, although one example from Lejre showed separation of silt grades. This was one of the distinguishing characteristics between 'panning'/infilling features and crusting features noted at Butser Ancient Farm; features with sorting were associated with near- or at-surface processes. Although internal slaking with sorting at depth (as seen in some cases under modern tillage) cannot be ruled out as influential, no examples of this type of pedofeature were seen in the experimental or archaeological samples. Further research examining both types of features within several soil types would aid greatly in this regard.
- A location that appears to echo or infill linear void spaces marking tillage feature cuts. These features were found inside (at the very base of the implement mark fill), at the cut line itself, and outside (immediately underneath or to the sides of the cut, or within shear planes related to implement action).

These infilling features were seen nowhere else in the implement mark features created, and it was possible to distinguish them from crusts also formed in one set of soils studied at Butser Ancient Farm. These types of pedofeatures have been found related to archaeological implement marks (e.g. Fisher & Macphail 1985, 98; Macphail 1987a; 1990a; 1992a; 1992b; French & Marsh 1999; French 2003; Lewis 2002). Other soil types may create variations on the theme, such as the placic layers lining ard mark cuts at Bjerre Site 7.

In all of these cases, these features were distinctive in thin section and very obviously echoing or at the feature cut line. They appear not to relate to rooting or earthworm action at the cut edges (*sensu* Atkinson 1957, 225), based on lack of positive evidence for either at this location, and are akin to pore infillings (Bullock *et al.* 1985) in their nature. However, other archaeological examples do show rooting following the voids marking implement feature cuts, including both living roots and root pseudomorphs; e.g. ancient spade mark cuts at Wandlebury, England, were marked by void space partly filled with calcium carbonate-replaced root pseudomorphs (Lewis 2003). This suggests that it is not only illuviation of fine material that preserves these boundaries, but that the void space created by the implement itself is somehow preserved and available for later root use, *i.e.* that the micro-compaction and planar void creation by the tilling tool is of primary importance for feature preservation over the long term. Care must be taken in interpreting tillage from the general occurrence of planar voids with infillings of fine materials; such features are also found, for example, in seasonally frozen soils that heave (Mermut and St. Arnaud 1981; Kemp *et al.* 1993). However, the arrangement and orientation of these, and the presence of other features (such as vesicles), make them distinct from the features studied here.

Regarding frequency of occurrence, or the likelihood of finding such pedofeatures during normal site sampling procedures, experimental sites with clear tillage marks show some variation on these structural (porosity) zones and associated pedofeatures to be ubiquitous, occurring in every experimental mark examined, as well as in many archaeological cases (see Lewis 1998). It is especially encouraging to note that at the Butser Ancient Farm site, where no macroscopically-visible marks were seen and where there was a great deal of faunal reworking, similar features were also located in the Ap horizon at depths which could be appropriate for the tillage implements and methods use there (but see caveat above re. using depth). These features were distinct in texture, location and orientation from crusting and other illuvial pedofeatures seen in the many thin sections examined from this site. As the Butser samples were taken randomly with regard to implement marks (since none were visible in the field), it would seem possible that random, but intensive, sampling of suspected or known buried horizons might produce similar results (e.g. French 2003). This level of study is rarely seen in archaeological fieldwork, however, where it is more likely that one or two samples would be taken from an old or buried soil profile for study of ancient land use (e.g. French & Lewis 2001).

CHARACTERISING IMPLEMENT MARKS MICROMORPHOLOGICALLY

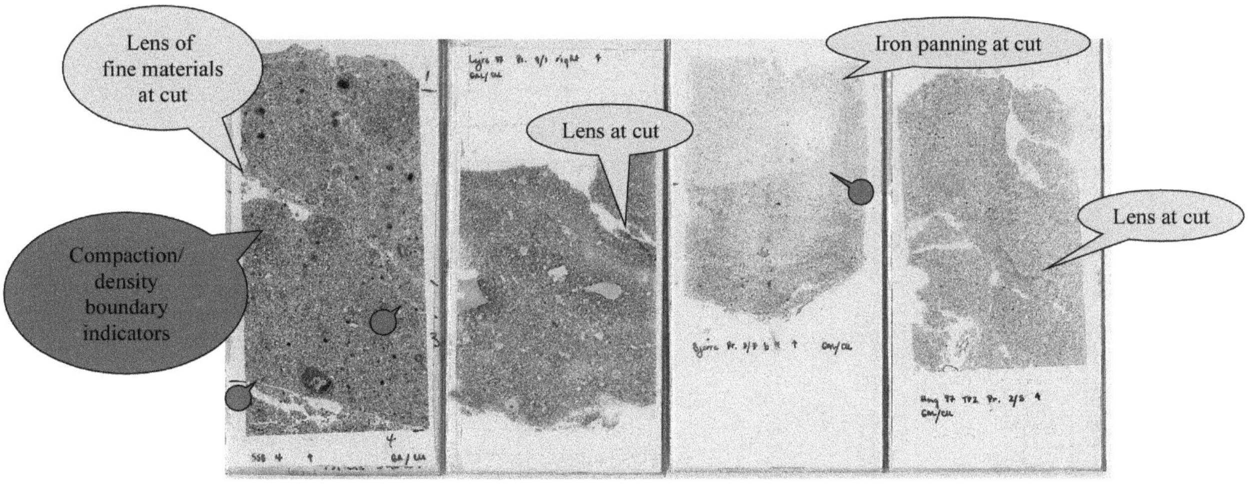

1. Experimental characteristics (lab)
2. Experimental characteristics (field)
3. Characteristics of very clear archaeological features
4. Characteristics of somewhat less clear archaeological features

Figure 17 Linking experiments to archaeology: zones of tillage marks; all thin sections were scanned 1:1 and are c. 6 x 12cm. This figure shows the process taken in this study to link experimental and archaeological features: 1. Silsoe soil bin experimental ard mark characterisation; 2. Lejre experimental ard mark showing the influence of real soil processes; 3. Bjerre Bronze Age ard mark showing characteristics of well-preserved features; 4. Iron Age/historic period implement mark in a spade-tilled horizon at Hengistbury Head Site 6 in England (Lewis 2002), where marks were less well-preserved.

The previously published record of thin sections related to tillage has produced features of the type discussed (see Chapters 2-3), although not very many of them. The results of this study suggest that this is not because they occur too rarely to be useful in cultivation identification, but that the record probably reflects sampling strategy more than either occurrence or survivability of such textural pedofeatures, although variation in soil types remains to be assessed in this regard. Interpreting ancient land use on a site from one or two soil thin sections may be the norm due to cost implications, but this sampling strategy can easily misrepresent land-use history. Where possible, and especially in research-driven situations, it is recommended that the practice of systematic and even repetitive sampling should be the norm for looking at land-use history from soils. In this regard, it should also be noted that the addition of other approaches (such as particle size analysis) can also augment the study of ancient tillage on a location-specific basis; this project was limited in methods due to the focus of its research aims.

Clay infillings and coatings

While 'dusty' clay coatings have been demonstrated to occur under certain conditions with tillage (Jongerius 1983), they have also been found in many disturbed soils under bare surfaces, regardless of whether these have been cultivated or not. For instance, raindrop splash on bare earth where vegetation has been removed by fire, making it susceptible to leaching, has been seen to create such features both experimentally and archaeologically (Boulbin 1976; Limbrey 1975; Gebhardt 1990, 33; 1993, 334-35; Macphail and Goldberg 1990; Macphail *et al.* 1990a, 54-55; Mikkelsen and Langohr 1996). In addition, such features have been related archaeologically to deforestation followed by illuviation (French 2005), and to tree throw disturbance (Macphail 1987d; Macphail *et al.* 1990a; Macphail and Goldberg 1990; Gebhardt 1993, 336). Goldberg (1992, 150-153) found both 'dusty' and limpid clay coatings at Berkhat Ram, Israel, dating to the Palaeolithic period, the former possibly relating to colluvial activity, and at Pech-de-l'Azé II rockshelter (France) 'dusty' clay coatings were associated with slope movement such as a mud slurry or solifluction (Courty *et al.* 1989; Goldberg 1992, 163). Irrigation and flooding have been seen to produce an abundance of clayey silt intercalations and coatings, although the combination of these with other features such as vesicles and certain types of crust formation may distinguish these processes from tilling archaeologically (Courty *et al.* 1996, 263; 1989, 137). These features cannot be linked to a sole cause (tillage) even in modern studies, and ancient examples must be interpreted cautiously, and in a location-specific manner that takes into

Table 33 Field descriptions of Profiles 2-5 from Bjerre Site 7 (Lewis 1998)

Horizon	Profile 2 - Area 76, outside house	Profile 3 - Area 76, inside house	Profiles 4 & 5 - Trench 29	Samples
Ap	0-10 cm	[13] - upper vegetation horizon, 0-10 cm.	n/a	no samples taken
		[14] - yellow drift sand, 10-13 cm.		no samples taken
Bfe/Bs	10-20 cm	[15] - brownish red/yellow sand, 13-35 cm.		3/1, 3/2
bAh (s)	20-25 cm	[5/6] - peaty layer, thinner vegetation layers below in some places, 35-40 cm.		3/1, 3/2, 3/3
Ea *(albic)*	25-30 cm	[7] - white out-washed sand, 40-46 cm.		3/3, 3/4a
Bfe/Bs *(spodic)*	30-70 cm - build up against house mound?	[8] - aeolian sand, rusty red becoming yellow with depth, 46-55 cm.		3/4b, 3/5
bAp_1 *(some possible spodic traits)*	[9] - greyish-yellow sand with rare dispersed cultural remains, presumably farmed, 70-80 cm (10 less than inside).	[9] - greyish-yellow sand with rare dispersed cultural remains, presumably farmed, 55-75 cm. Undulating boundary with [1a].	[4] - ploughed upper part of culture layer mixed with sand, 2-3 cm depth from exposed level (c. 75 cm depth).	2/1, 3/5, 3/6, 3/7a, 3/7b, 4/1, 5/1
➔	ard marks	ard marks	ard marks	4/1, 5/1
'culture layer' *(some possible spodic traits)*	[1a] - brownish grey sand with charcoal flecking, becoming less sandy with depth, a brown line at the bottom *(possibly placic? lamina?)*, 80-85 cm. Layer becomes less distinct here and is hard to tell from [2].	[1a] – brownish grey sand with charcoal flecking, becoming less sandy with depth, with a 1 cm thick dark brown organic lens at the base *(possible lamina associated with podzolisation?)*, 75-80 cm.	[3] - more steel-grey colour than normal culture layer, 75-86 cm.	2/1, 2/2, 3/7b, 3/8, 4/1, 5/1
bAp_2 *(some possible spodic traits)*	[2] - uniform grey sand. Old plough layer, 85-95 cm.	[2] – uniform grey sand. Old plough layer, 80-90 cm.	[1] - grey plough layer under culture layer, 86-93 cm.	2/2, 2/3, 2/4, 3/9
C		[3] – white sterile sand subsoil, 90-93 cm exposed.		no samples taken
	[19] - light grey sand with rare dispersed charcoal flecks, 95-100 cm.		[18] - as in Profile 2 [19], 93-95 cm exposed.	2/4, 2/5
C	[20] - white sand layers alternating with greyish-brown sandy layers, 100-110 cm.			2/5

account soil and environmental, as well as archaeological, influences. Where soil constituents and climate are favourable to clay translocation and illuviation (Nettleton *et al.* 1987, 38), a physical disturbance such as that created by tillage should result in bond and aggregate disruption which should create 'dusty', 'dirty' or 'silty' clay textural features. Such soils should possess 'dusty' clay when disturbed and these features should thus be expected to be regularly encountered in these soils when they have been cultivated. If such a soil does not possess these features, it may seem reasonable to suggest that it has not been disturbed in any major way (and therefore not tilled). However, there is a chance that some textural features are lost to or masked in specific horizons or profiles through biotic activity, leaching, etc, as discussed above. For instance, at the Bronze Age field system at Trethellan Farm, England, Macphail (1989) suggests that strong post-cultivation earthworm activity reworked an earlier slaked fabric (with very 'dusty' clay pedofeatures) created during cultivation or trampling, to produce a biological fabric. It is also very important to note that the creation of such features depends on whether or not a soil has been decalcified (Duchaufour 1982). Gebhardt suggests (1995, 34) that no 'dusty' clay coatings were produced during her experiments in tillage because the soils were not sufficiently decalcified. This may also have had an impact on the Little Butser samples discussed above. It has been noted by Romans and Robertson (1983b, 64) that clay mobilisation may not be affected by cultivation if there is enough organic matter to stabilise the soil, that clay may be 'recycled' in a soil with an active earthworm population, and that there may be secondary destruction of clay in surface horizons.

The summarized findings suggest that the idea of a compaction zone at the base of tillage, which seems generally accepted in archaeological studies, should be expanded to include the sides of implement marks as a location in which textural and amorphous pedofeature accumulation occurs. Regarding illuviation, it appears most important that the zone under the implement mark cut be compacted relative to the overlying material (fill of the mark/Ap horizon) and not compacted in some general sense – the 'denser zones' under the ard mark cuts at Bjerre, for instance, are composed of loose sand (Lewis 1998). While they have a slightly lower porosity than do the fills, much of this reduction may be related to later infilling. Nevertheless, the cuts of features were the location at which coatings accumulated, suggesting that even a minor change in density can produce a major effect, such as panning. The chemical variations also implied through pan formation have not been studied here, but must also play a major role in the development and preservation of these zones.

It is possible that these boundaries have a role in later horizonation. Carrying on here with the example of podzolising soils, placic (iron, iron-manganese or iron-amorphous organic matter panning) horizons have been noted to form at the base of cultivation. Proudfoot (1958, 193-195), for instance, discusses iron pan formation at the base of historic rig cultivation (spade dug) at Torr, Northern Ireland, and remarks that the depth of Neolithic cultivation at Goodland, Northern Ireland, was the location in the profile for later iron pan formation. At Torr, the pan divided a zone of looser topsoil from a compact zone at the base of tilling (*ibid*, 193-194). At Dark Hat, England, Reynolds and Catt (1987) suggest that an iron pan horizon developed at the base of a cultivation compaction zone. At Bjerre, in addition to the iron accumulation features seen at ard mark edges discussed above, marking a thin placic horizon development in the profile, there was also a placic horizon proper seen in Profile 1 (Lewis 1998). This, however, formed within and/or underneath the base of a buried sandy horizon (context [11]), not a known ancient cultivation soil, and does not appear, therefore, to mark a profile change associated with tillage. As in all cases discussed here, caution in interpretation appears to be advisable, and in-depth study of the specific soils and archaeological contexts in question is required before linking soil features directly to ancient cultural activity.

Distinguishing implement types through study of feature characteristics

Despite a long focus on factors such as the size and shape of fields, distances between implement marks, and depth of tillage, the most obvious means of distinguishing between ard ploughing and hoe tillage *versus* mouldboard ploughing and spade tillage is through the investigation of indicators of horizon reversal (inversion). This volume does not focus on this issue, because no implement-focused experiments exploring the features of turning implements were conducted. Instances of horizon reversal created by turning implements have been seen in the field and in thin section in both experimental and archaeological studies (e.g. Macphail 1994a), and have been discussed in relation to the work described here in Lewis (2002); this work is relevant to the discussion above, as the various zones defined in association with tillage features, and in particular the boundary created by the cut zone, are important in regard to distinguishing between general implement types.

There are three main zones within tillage features to examine both macroscopically and microscopically for inversion: the main fill, the lowermost (stratigraphically) fill, and the cut itself. Turning implements have been seen to lead to the creation of archaeological features whose main fills show enhanced organic matter contents in comparison to surrounding contexts, including the contexts from which they were created. Both plough mark and spade mark fills at Hengistbury Head Site 6, England (Lewis 2002) were distinguished by a more organic nature than the material under- and overlying them. Indeed, the colouring associated with this was the factor that made them visible in the field in the first place. This enhanced organic nature presumably relates to the turning in of upper A horizons (litter/stubble) within the marks only. In addition, as discussed previously in this volume, field studies show the inclusion of inverted material in cases where turning implements have interacted with more than one horizon, and that these result in feature fill arrangements that relate to implement action.

This may distinguish plough marks and spade marks from ard marks in some circumstances. In certain ard-mark tilled horizons, such as at Bjerre (Lewis 1998) a more organic nature was noted as a characteristic of the cultivation horizons as a whole, but the ard marks themselves were filled with materials obviously originating from specific horizons, either those immediately above or below, except in the case of *in situ* alteration, such as leaching. Ard tillage may lead to a more organic horizon, as opposed to simply enriched feature fills, because it mixes over wide area *vs.* just within features. The organic Ap horizons seen with mouldboard ploughing historically may also reflect multiple tilling events and the creation of ridges. This appears to be the case, for instance, with the mouldboard plough mark interpreted by Macphail (1998a; Goldberg and Macphail 2006, 206), where 'anthropogenic inclusions' were not limited to the plough mark fills, and there were indications of multiple tilling events. As with all of the characteristics discussed in this volume, the specific circumstances of each site over-ride any generalisations made from experimental archaeology and generalised pedogenetic modelling of the impact of cultural activity.

Table 34 Summary of the micromorphological characteristics of ard marks at Bjerre Site 7 (Lewis 1998; percentages are visual estimates of area)

	Profile	
Zone	BJE 4 & 5	BJE 3
FURROW FILL (Ap)	[4] upper (bAp) – single grain, 25% porosity (packing pores); <5% organic matter; gefuric; black fine groundmass CUT of the upper furrow.	[9] (bAp) – single grain; 25% porosity (packing pores); 5-10% organic matter; gefuric-enaulic; fine groundmass - reddish-brown organo-mineral coatings
FURROW FILL (basal)-CUT ZONE (1-2 cm thick)	[4] lower – single grain, intergrain microaggregate; 20-25% porosity (packing pores); 10-20% organic matter; chitonic-gefuric; black fine groundmass; V-shaped lenses of densely packed fines; occasional planar voids and lines of connecting pores	[1a]upper – single to bridged grain; 20% porosity (packing pores); 10% organic matter; chitonic; clay-organo-mineral accumulation zone running up sides and along base of furrow AMORPHOUS ACCUMULATION
CUT/BASE	OF AP HORIZON	
(<1 cm thick)	[3] upper – single grain, intergrain microaggregate; 20% porosity (packing pores); 20-25% organic matter; chitonic-gefuric; rare black coatings CLAY ACCUMULATION	[1a] m – bridged grain; 20% porosity (packing pores); 15-20% organic matter; chitonic; very dark reddish-brown-black amorphous organic accumulation; limpid and dusty clay accumulation; undulating horizontally-oriented lenses of dense amorphous organic matter (possibly infilling shear planes?) AMORPHOUS AND CLAY ACCUMULATION
	[3] middle – single grain, intergrain microaggregate; 30-35% porosity (packing pores); 15-20% organic matter; chitonic-gefuric; black coatings (rare)	[1a] l – singles grain, pellicular grain; 15-20% porosity (packing pores); 30% organic matter; chitonic-enaulic; black amorphous organic in increased proportions. 'CULTURE LAYER'
	[3] lower – single grain, intergrain microaggregate; 30-35% porosity (packing pores); 25-30% organic matter; chitonic-gefuric; rare infillings	

Next to the cut, on the base and at one side, a lens of upper topsoil may be found in mouldboard plough marks, attesting to the turning over of the very upper horizon into the base of the feature. In addition, in the void space(s) marking the cut of the tillage disturbance zone, a thin band or discontinuous presence of plant remains may be encountered. While on some occasions these plant remains might indicate root penetration into a zone loosened by the action of the implements, the association of these remains in this position (at the cut) in samples known to have been spade dug or mouldboard ploughed, and the general absence of such remains in the same location in ard-tilled samples suggests that they may indeed be related to the turning action associated with mouldboard ploughs and spades. Ards do not invert, but push the uppermost layer to the sides – features showing turning characteristics should thus not normally be seen as a product of ard tillage. In some occasions, the inversion of upper horizons is very clear in the field, producing an obvious organic layer constituting the basal stratigraphic fill of spade and mouldboard plough marks. Using these indicators it has proven possible to suggest that the interpreted ard marks associated with the upper cultivation soil at Hengistbury Head Site 6 may actually be more indicative of the disturbance patterns created by mouldboard ploughs (Lewis 2002).

The presence of large aggregates or a layer of material from lower horizons in the uppermost part of a tillage feature should likewise lead to the interpretation of the use of a turning implement. The frequent location of macroscopically small patches of sand in the tops of spade marks at Hengistbury Head Site 6 (Lewis 2002) is a clear example of such inversion (also see Gebhardt 1995). At Lejre, ard tilling was seen to bring Bt horizon fragments into the A horizon, but these were mainly near the base of the topsoil (Lewis 1998).

The inversion patterns described here are not visible in all situations. At Butser Ancient Farm, for example, spade-dug fields produced almost no macroscopically-visible tillage marks, and the only possible surviving indicators of inversion were some crust fragments at depth, but these could have arrived at their location in the profile by other means.

As noted in Chapter 4, the presence of 'rolled' and flattened peds at the cut boundary may be related to the action of traction implements; this was seen in two instances in the experimental study; further investigation would be required to determine whether these features also relate to mouldboard ploughs, and whether they can survive archaeologically.

Summary

The most distinctive micromorphological features of tilling seen are those associated with implement feature cuts. The cut represents a structural and textural boundary within the soil profile, and appears to retain its distinguishing characteristics over long periods of time in many instances. The cuts of tillage features examined were always marked microscopically by at least two characteristics: aligned pore space along the cut itself (planar voids, shear planes), and a density change between the 'inside' and 'outside' of the feature (loosening inside, compaction outside). Other characteristics include associated distortion under the base and sides of the cut, and an accumulation zone for fine soil particles at the base of the cut itself, or immediately outside the cut (usually within 1-2cm). These zones are both created by the action of tilling itself, and secondary infilling of spaces created by the implement. Archaeological examples of these features are regularly, although not frequently, found through soil micromorphological study. Variations on this theme include accumulation zones related to leaching and podzolisation, rooting zones attested to by pseudomorphs found in feature cuts, differences in organic matter content, 'bleaching' and re-located fragments of upper or lower horizons between the fills (inside) and surrounding soil (outside) (Lewis 1998). All of these characteristics of archaeological cultivation features can be directly related to the original action of tillage, but must be interpreted on a soil-, site- and environment-specific basis.

Chapter 7 Conclusion

The results of this study suggest that the combination of experimental field and laboratory work, along with examples from archaeological sites both with and without clear tillage indicators at a macroscopic scale (Lewis 1998; 2002), should continue to be pursued in the search for a better understanding and characterisation of microscopic tillage morphologies.

It is very important that sites with good preservation be further investigated on both a macroscopic and microscopic basis. The study of known archaeological features in comparison to those created experimentally has advanced our understanding of tilling features and post-depositional profile and horizon characteristics related to cultivation. A focus on archaeological features themselves shows potential both for enhancing our understanding of micromorphological indicators which may be of use in identifying tillage where no such features exist, and also holds potential for studies of implement type and tillage regime.

Although base-line methodological research into tillage indicators will continue to require a research-oriented excavation approach, it is expected that most data sources will arise from archaeological contract situations. Tillage features are central to the understanding of ancient arable practices. The presence and connotations of arable activities are issues of great significance in archaeology, and any archaeological evidence of these activities which can be integrated spatially and temporally with other categories of features should be explored, or at least fully recorded, with this in mind.

Features such as ard and spade marks, along with potentially tilled ancient soil horizons should be investigated in more detail in the field and, where possible, through post-excavation analyses, including soil micromorphological studies. Many of the microscopic features identified in thin section could be located in the field if more attention were paid to the detailed recording and contextual interpretation of soil horizons and the internal stratigraphy of archaeological features. The genetic relationship of such features to specific processes, such as tilling, may be difficult without microscopic approaches. For instance, it is possible to see and record very fine layers distinguished by texture and colour in the field, but relating these features to processes created by ard or plough tilling may be more complicated without more detailed, and site-specific, study. This research suggests that soil micromorphology should not be used as a replacement for much-needed detailed recording of contexts in the field, but that is has an enormously important role to play in archaeological interpretation and characterisation of soil features and horizons.

One of the main difficulties in studying ancient cultivation from the medium in which it occurred – the soil – is that relatively few archaeological soils are reported in enough detail to be assessed for particular cultural activities such as tillage. The lack of detailed published information of most relevant features from archaeological sites with tillage marks was a major problem noted by Lewis (1998); beyond information on feature presence, and plan and (occasionally) sections of implement mark features, most investigators have not published sufficient ,information on fills, cuts or relationships to soil horizons and later activities to enable discussion of the production, survival, extent or, most important, meaning of tillage marks. There are notable exceptions to this, and many of these cases have been cited throughout this volume. Without more extensive reporting it will continue to remain difficult to link ancient land-use practices to cultural change on specific sites and in archaeological landscapes.

The results of this experimental study suggest that soil micromorphology has a great deal of potential for characterizing ancient features, and relating these to soil profile and horizon characteristics. There are types of pedofeatures and structural and depositional boundaries that appear to mark the base of cultivation, both within implement features and within tilled soil horizons, and these are found archaeologically as well as experimentally. This supports the idea that the method has potential for not only characterising, but also identifying ancient tillage where site conditions are appropriate, on a site-by-site basis that takes into account both local soil histories and archaeological taphonomic issues. In addition, it appears to be possible to archaeologically distinguish implement type (Lewis 2002), although this may also be restricted to particular preservation environments, and requires further investigation.

Regarding more specific issues, this study disagrees with some deeply-held beliefs regarding the impact of non-mechanised tillage on the soil. Ard tillage appears to create compaction/compression zones, even in the absence of trampling, and these zones appear to play a role in later soil horizon formation processes in certain environments, *i.e.* they have a long-lasting impact on the soil. Ancient cultural activities are well known in archaeology to have a strong impact on modern soil and sediment characteristics; these are certainly not as influential as modern mechanized farming methods, but it seems that even low- or 'no'-till options can be implicated in long-term pedogenetic, and thus environmental changes.

The suggestion that textural pedofeatures cannot generally be used to distinguish specific processes such as tillage may be somewhat premature. The search for specific causes may be difficult from an archaeological perspective, given the taphonomic issues, not least of truncation, but in this experimental study textural and organizational differences between features such as surface crusts, infillings at the base of tillage, 'dusty' clay coatings, whole-soil coatings and panning features were distinguishable (Lewis 1998). Since these micro-features develop at different locations in features and soil profiles, this study has an optimistic conclusion: with further study involving different soil types, it seems logistically and logically possible to use these features to discuss at least the presence or absence of ancient tillage, if not the specific implements involved. Soil micromorphology, while a complicated way to gather evidence, is an excellent method for understanding how features are created, how features are altered over time, and how they themselves alter surrounding contexts and environments. The resource of cultural data held in ancient soils and sediments is sorely underexplored as evidence in landscape archaeology, and has not seen the substantial theoretical discourse of a type common to other lines of archaeological evidence. Perhaps this is because soil is often seen as a background context in archaeology, as opposed to a feature created by cultural activity. In the case of ancient arable agriculture, cultivated soils are the most widespread artefact relating to land use, and all approaches that can help us to interpret its archaeological significance will enable a better understanding of the development, impact and meaning of arable farming in the past.

Finally, the combination of experimental archaeology with the logic of soil pedogenetic theory meant that it was possible in this study to take a series of reasonable inferential steps through the creation, preservation, alteration and localised impact of particular types of archaeological tillage features. While the focus of many researchers studying ancient agriculture understandably targets the relationship of land-use or other agricultural indicators to wider archaeological, environmental and social issues, those of us using soil micromorphology as an archaeological data source need to also continue to address the fundamental methodological and interpretative issues of this complex but highly rewarding line of evidence. Soil micromorphology is one of the few archaeological approaches that directly addresses and investigates the medium in which cultivation occurs, and explores how ancient cultural activity interacts with past and present-day soils and sediments. It is a primary tool in the study of context and taphonomy, but requires site-specific interpretation due to the nature of the material studied. Over time and with continued study, a database that can adequately be interpreted on a regional or general interpretative basis can be developed, with which to address other proxy indicators and the ethnographically-based models of the social history and significance of ancient arable farming/ The case of the many examples cited in this volume from research in the UK in particular, shows how soil micromorphological data can become involved in large-scale understandings of environmental and cultural history. Soils offer us the opportunity to explore direct cultural interaction with the earth, in an environmentally, spatially and temporally sensitive medium, and with further investigation and exploration these can provide us with an alternative perspective from which to view archaeological landscapes and ancient cultural activities.

Bibliography

Aaby, B. 1993. Man and the environment. In Hvass, S. and Storgaard, B. (eds.) *Digging into the Past: 25 Years of Archaeology in Denmark*, 16-19. Copenhagen and Hojbjerg: Royal Society of Northern Antiquaries and Jutland Archaeological Society.

Aberg, F.A. and Bowen, H.C. 1960. Ploughing experiments with a reconstructed Donneruplund ard. *Antiquity* 34, 144-47.

Adams, M. 1996. Iron Age ridge and furrow? So it seems. *British Archaeology* 13, 6.

Adderley, W.P., Simpson, I.A. and Davidson, D.A. 2006. Historic landscape management: a validation of quantitative soil thin-section analyses. *Journal of Archaeological Science* 33(3), 320-34.

Addiscott, T.M. 1993. Simulation modelling and soil behaviour. *Geoderma* 60, 15-40.

Aguilar, J. Fernández, J., Ortega, E., de Haro, S. and Rodríguez, T. 1990. Micromorphological characteristics of soils producing olives under nonploughing compared with traditional tillage methods. In Douglas, L.A. (ed.) *Soil Micromorphology: A Basic and Applied Science. Proceedings of the 8th International Working Meeting of Soil Micromorphology, San Antonio, Texas - July 1988*, 25-32. Developments in Soil Science 19. Amsterdam: Elsevier Science Publishers B.V.

Aguilar, J., Guardiola, J.L., Barahona, E., Dorronsoro, C. and Santos, F. 1983. Clay illuviation in calcareous soils. In Bullock, P. and Murphy, C.P. (eds.) *Soil Micromorphology, Volume 2: Soil Genesis. Proceedings of the 6th International Working Meeting on Soil Micromorphology, London 1981,* 541-50. Berkhamsted: AB Academic Publishers.

Alexandrovskiy, A.L. 1996. Application of pedology methods to archaeology – results of a long-term research programme 1970-1990. *Proceedings of the 6th Nordic Conference on the Application of Scientific Methods in Archaeology, Esbjerg, 1993*, 323-35. Arkæologiske Rapporter fra Esbjerg Museum, No. 1. Esbjerg: Esbjerg Museum.

Alexandrovskiy, A.L. and Gol'yeva, A.A. 1997. Palaeosols of the Scythian time: phytolithic studies. In Pinilla, A., Juan-Tresserras, J. and Machado, M.J. (eds.) *Estado Actual de los Estudios de Fitolitos en Suelos y Plantas*, 267-73. Consejo Superior de Investigaciones Científicas Monografias 4. Madrid: Centro de Ciencias Medoambientales del Consjo Superior de Investigaciones Científicas.

Allen, M.J. 1992. Products of erosion and the prehistoric land-use of the Wessex chalk. In Bell, M. and Boardman, J. (eds.) *Past and Present Soil Erosion*, 37-52. Oxbow Monograph 22. Oxford: Oxbow Books.

Allen, M.J. 1994. *The Landscape History of the Southern English Chalkland with an Evaluation of the Beaker Period using Environmental and Cultural Indicators.* Unpublished Ph.D. thesis, University of Southampton.

Allen, T.G. and Robinson, M.A. 1993. *The Prehistoric and Iron Age Enclosed Settlement at Mingies Ditch; Hardwick-with-Yelford,,Oxon.* Thames Valley Landscapes: the Windrush Valley Volume 2. Oxford: Oxford Archaeological Unit/Oxford University Committee for Archaeology.

Allen, M.J., Bloomfield, C., Macphail, R.I. and Wyles, S.F. 1992. A record of soil and molluscan changes resulting from landuse management at Butser Archaeological Farm: some preliminary results. Report for *Butser Archaeological Yearbook.*

Altemüller, H.J. and Banse, H.J. 1964. Die Bedeutung der Mikromorphologie hinsichtlich der organischen Düngung. In Jongerius, A. (ed.) *Soil Micromorphology*, 467-76. Amsterdam: Elsevier.

Anderson, P.C. 1992. Experimental cultivation, harvest and threshing of wild cereals and their relevance for interpreting the use of Epipalaeolithic and Neolithic artefacts. In Anderson, P.C. (ed.) *Préhistoire de l'Agriculture: Nouvelles Approches Experimentales et Ethnographiques*, 179-209. Monographie du CRA No. 6. Editions due CNRS. Paris: Centre Nationale de la Recherche Scientifique.

Andersen, S.T. 1989. Natural and cultural landscapes since the Ice Age. Shown by pollen analyses from small hollows in a forested area in Denmark. *Journal of Danish Archaeology* 8, 188-99.

Andersen, S.T. 1993. Early agriculture. In Hvass, S. and Storgaard, B. (eds.) *Digging Into the Past: 25 Years of Archaeology in Denmark*, 88-91. Copenhagen and Hojbjerg: Royal Society of Northern Antiquaries and Jutland Archaeological Society.

Andreini, M.S. and Steenhuis, T.S. 1990. Preferential paths of flow under conventional and conservation tillage. *Geoderma* 46, 85-102.

Arroyo-Kalin, M. 2008. *Steps towards an ecology of landscape: a geoarchaeological approach to the study of anthropogenic dark earths in the central Amazon region, Brazil*. Unpublished PhD thesis. University of Cambridge.

A.S.A. (American Society of Agronomy and the Soil Science Society of America). 1982. *Predicting Tillage Effects on Soil Physical Processes and Properties*. American Society of Agronomy Special Publication Number 44. Madison: American Society of Agronomy and the Soil Science Society of America.

Ashbee, P., Cameron, A., Davies, D.M., Ellis, C., Evans, J., Grigson, C., Haddon-Reece, D., Meddens, B. and Shaw, G. 1986. The excavation of Milton Lilbourne barrows 1-5. *Wiltshire Archaeological and Natural History Magazine* 80, 23-96.

Ashbee, P., Smith, I.F. and Evans, J.G. 1979. Excavation of three long barrows near Avebury, Wiltshire. *Proceedings of the Prehistoric Society* 45, 207-300.

Asplund, H., Formisto, T. and Illmer, K. 1989. Kotirinne – a late neolithic mixed farming site: osteological and chemical investigations at the Kotirinne dwelling site at Niuskala, Turku, SW Finland. *Norwegian Archaeological Review* 22(2), 119-29.

Atkinson, R.J.C. 1957. Worms and weathering. *Antiquity* 31, 219-33

Avery, B.W. 1980. *The Soil Classification of England and Wales*. Soil Survey Technical Monograph 14. Harpenden: Soil Survey.

Bakels, C.C. 1988. Pollen from plaggen soils in the province of North Brabant, The Netherlands. In Groenman-van Waateringe, W. and Robinson, M. (eds.) *Man-made Soils*, 35-54. British Archaeological Reports International Series 410. Oxford: British Archaeological Reports.

Barker, G. 1985. *Prehistoric Farming in Europe*. Cambridge: Cambridge University Press.

Barratt, B.C. 1968. Micromorphological observations on the effects of land use differences on some New Zealand soils. *New Zealand Journal of Agricultural Research* 11(1), 101-30. Soil Bureau Publication 408.

Barratt, B.C. 1971. A micromorphological investigation of structural changes in the top soil of Patamahoe clay used for market gardening. *New Zealand Journal of Science* 14(3), 580-98.

Barrett, J.C. 1994. *Fragments from Antiquity: An Archaeology of Social Life in Britain, 2900-1200 BC*. Oxford: Blackwell.

Bech, J.-H. 1993. Settlements on the raised sea-bed at Bjerre, northern Thy. In Hvass, S. and Storgaard, B. (eds.) *Digging into the Past: 25 Years of Archaeology in Denmark*, 142-43. Copenhagen and Hojbjerg: Royal Society of Northern Antiquaries and Jutland Archaeological Society.

Becker, C.J. 1992-3. Viggo Nielsen: *Jernalderens Pløjning. Store Vildmose* (Iron Age Ploughing - Store Vildmose) (Book review, translated by D. Liversage). *Journal of Danish Archaeology* 11, 165-67.

Beckman, G.G. and Smith, K.J. 1974. Micromorphological changes in surface soils following wetting, drying and trampling. In Rutherford, G.K. (ed.) *Soil Microscopy: Proceedings of the 4th International Working Group Meeting on Soil Micromorphology, Department of Geography, Queen's University, Kingston, Ontario, 27th-31st August 1973*, 832-45. Kingston: Limestone Press.

Behre, K.E. 1981. The interpretation of anthropogenic indicators in pollen diagrams. *Pollen et Spores* 23(2), 225-45.

Bell, M. 1977. Excavations at Bishopstone. *Sussex Archaeological Collections* 115, 1-299.

Bell, M. 1981a. Seaweed as a prehistoric resource. In Brothwell, D. and Dimbleby, G. (eds.) *Environmental Aspects of Coasts and Islands*, 117-26. British Archaeological Reports International Series 94. Oxford: British Archaeological Reports.

Bell, M. 1981b. Valley sediments and environmental change. In Jones, M. and Dimbleby, G. (eds.) *The Environment of Man: the Iron Age to the Anglo Saxon Period*, 75-91. British Archaeological Reports British Series 87. Oxford: British Archaeological Reports.

Bell, M. 1982. The effects of land use and climate on valley sedimentation. In Harding, A.F. (ed.) *Climatic Change in Later Prehistory*, 127-42. Edinburgh: Edinburgh University Press.

Bell, M. 1983. Valley sediments as evidence of prehistoric land-use on the South Downs. *Proceedings of the Prehistoric Society* 49, 119-50.

Bell, M. 1992. The prehistory of soil erosion. In Bell, M. and Boardman, J. (eds.) *Past and Present Soil Erosion: Archaeological and Geographical Perspectives,* 21-36. Oxbow Monograph 22. Oxford: Oxbow Books.

Bellomo, R.V. 1993. A methodological approach for identifying archaeological evidence of fire resulting from human activities. *Journal of Archaeological Science* 20, 525-53.

Bender, B. 1978. Gatherer-hunter to farmer: a social perspective. *World Archaeology* 10(2), 204-22.

Bennett, K.D., Boreham, S., Sharp, M.J. and Switsur, V.R. 1992. Holocene history of environment, vegetation and human settlement on Catta Ness, Lunnasting, Shetland. *Journal of Ecology* 80, 241-73.

Besnard, E., Chenu, C., Balesdent, J., Puget, P. and Arrouays, D. 1996. Fate of particulate organic matter in soil aggregates during cultivation. *European Journal of Soil Science* 47, 495-503.

Bethell, P.H. and Máté, I. 1989. The use of soil phosphate analysis in archaeology: a critique. In Henderson, J. (ed.) *Scientific Analysis in Archaeology*, 1-29. Oxford University Committee Monograph No. 19. Oxford.

Blackwell, P.S., Jayawardane, N.S., Green, T.W., Blackwell, J.T. and Beatty, H.J. 1991. Subsoil macropore space of a transitional red-brown earth after either deep tillage, gypsum or both. I. Physical effects and short-term changes. *Australian Journal of Soil Research* 29, 123-40.

Bobrova, E. and Bobrov, A. 1997. Phytoliths in soils: species composition, distribution along a soil profile, and value as environmental indicators. In Pinilla, A., Juan-Tresserras, J. and Machado, M.J. (eds.) *Estado Actual de los Estudios de Fitolitos en Suelos y Plantas*, 5-13. Consejo Superior de Investigaciones Científicas Monografias 4. Madrid: Centro de Ciencias Medoambientales del Consjo Superior de Investigaciones Científicas.

Bogucki, P. 1988. *Forest Farmers and Stockherders. Early Agriculture and its Consequences in North-central Europe.* Cambridge: Cambridge University Press.

Bogucki, P. 1996. Sustainable and unsustainable adaptations by early farming communities of northern Poland. *Journal of Anthropological Archaeology* 15, 289-311.

Boiffin, J. and Bresson, L.M. 1987. Dynamique de formation des croûtes superficielles: apport de l'analyse microscopique. In Fedoroff, N., Bresson, L.M. and Courty, M.A. (eds.) *Soil Micromorphology. Proceedings of the 7th International Working Meeting on Soil Micromorphology, Paris - July 1985*, 393-99. Plaisir: l'Association Française pour l'Étude du Sol.

Boismier, W.A. 1995. *Modelling the Effects of Tillage Processes on Artefact Distributions in the Ploughzone: A Simulation Study of Tillage-Induced Pattern Formation.* Unpublished PhD thesis. University of Cambridge.

Borchert, H. 1961. Einfluss der Bodenerosion auf die Bodenstruktur und Methoden zu ihrer Kennzeichnung. *Geologische Jahrbuch* 78, 439-502.

Borchert, H. 1964. Mikromorphologische Beobachtungen an meliorierten Böden. In Jongerius, A. (ed.) *Soil Micromorphology*, 459-65. Amsterdam: Elsevier.

Borchert, H. 1967. Untersuchungen der Bodengefügeveränderung von meliorierten Böden. *Geoderma* 1, 371-90.

Boserup, E. 1965. *The Conditions of Agricultural Growth: The Economics of Agrarian Change under Population Pressure.* Chicago: Aldine Publishing.

Boserup, E. 1970. *Woman's Role in Economic Development.* London: George Allen and Unwin.

Bouché, M.B. 1979. Contribution des lombriciens aux migrations d'éléments dans les sols tempérés. Migration organo-minérales dans les sols tempérés. *Colloques Internationaux du C.N.R.S.* 303, 145-53.

Boulbin, M. 1976. *Les Sols de la Fôret de Rennes et l'Influence des Incendies surs leurs Propriétés Morphologiques et Physico-chimiques.* Thèse de 3e cycle, Université de Rennes 1.

Bouma, J. 1969. *Microstructure and Stability of Two Sandy Loam Soils with Different Soil Management.* Unpublished thesis. Agricultural University of Wageningen.

Bouwman, A.F. 1989. Modelling soil organic matter decomposition and rainfall erosion in two tropical soils after forest clearing for permanent agriculture. *Land Degradation and Rehabilitation* 1, 125-40.

Bowen, H.C. 1963. *Ancient Fields: A Tentative Analysis of Vanishing Earthworks and Landscapes.* London: British Association for the Advancement of Science.

Bowen, H.C., Evans, J.C. and Race, E. 1978. An investigation of the Wessex linear ditch system. In Bowen, H.C. and Fowler, P.J. (eds.) *Early Land Allotment*, 149-53. British Archaeological Reports, British Series 48. Oxford: British Archaeological Reports.

Boyd, W.E. 1982a. Subsurface formation of charcoal and its possible relevance to the interpretation of charcoal remains in peat. *Quaternary Newsletter* 38, 15-16.

Boyd, W.E. 1982b. Subsurface formation of charcoal: an unexplained event in peat. *Quaternary Newsletter* 38, 15-16.

Bradley, R. 1972. Prehistorians and pastoralists in Neolithic and Bronze Age England. *World Archaeology* 4, 192-204.

Bradley, R. 1978a. Prehistoric field systems in Britain and north-west Europe – a review of some recent work. *World Archaeology* 9(3), 265-80.

Bradley, R. 1978b. *The Prehistoric Settlement of Britain.* London: Routledge, Kegan and Paul.

Bradley, R. 1992. The gravels and British prehistory from the Neolithic to the Early Iron Age. In Fulford, M. and Nichols, E. (eds.) *Developing Landscapes of Lowland Britain. The Archaeology of the British Gravels: A Review*, 15-22. London: Society of Antiquaries of London.

Brady, N.C. 1974. *The Nature and Properties of Soils.* 8th edition. New York: Macmillan Publishing Co. Ltd.

Bratanic, B. 1963. A note on the typology of ploughing implements. *VIe Congrès International des Sciences Anthropologiques et Ethnographiques, Paris 1960*, 511-16.

Bray, F. 1986. *The Rice Economies: Technology and Development in Asia.* New York: Blackwell.

Bresson, L.-M. and Boiffin, J. 1990. Morphological characterization of soil crust development stages on an experimental field. *Geoderma* 47, 301-25.

Bridges, E.M. 1978. *World Soils.* 2nd edition. Cambridge: Cambridge University Press.

Brochier, J.E. 1993. Çayönü Tepesi. Domestication, rythmes et environnement au PPNB. *Paléorient* 19(2), 39-49.

Brochier, J.E., Villa, P., Giacomarra, M. and Tagliacozzo, A. 1992. Shepherds and sediments: geo-ethnoarchaeology of pastoral sites. *Journal of Anthropological Archaeology* 11, 47-102.

Brown, T. 1997. Clearances and clearings: deforestation in Mesolithic/Neolithic Britain. *Oxford Journal of Archaeology* 16(2), 133-46.

Bryant, R.G. and Davidson, D.A. 1996. The use of image analysis in the micromorphological study of old cultivated soils: an evaluation based on soils from the island of Papa Stour, Shetland. *Journal of Archaeological Science* 23, 811-22.

Bullock, P. 1983. The changing face of soil micromorphology. In Bullock, P. and Murphy, C.P. (eds.) *Soil Micromorphology, Volume 1: Techniques and Applications. Proceedings of the 6th International Working Meeting on Soil Micromorphology, London 1981*, 1-18. Berkhamsted: AB Academic Publishers.

Bullock, P. and Murphy, C.P. 1979. Evolution of a palaeo-argillic brown earth (Paleudalf) from Oxfordshire, England. *Geoderma* 22(3), 225-52.

Bullock, P., Federoff, N., Jongerius, A., Stoops, G.J. and Tursina, T. 1985. *Handbook for Soil Thin Section Description.* Wolverhampton: Waine Research Publishers.

Bunney, S. 1990. Prehistoric farming caused devastating soil erosion. *New Scientist* 125(1705), 20.

Butler, A. 1992. Pulse agronomy: traditional systems and implications for early archaeology. In Anderson, P.C. (ed.) *Préhistoire de l'Agriculture: Nouvelles Approches Experimentales et Ethnographiques*, 67-78. Monographie du CRA No. 6. Editions du CNRS. Paris: Centre Nationale de la Recherche Scientifique.

Butler, S. 1995. Post-processual palynology. *Scottish Archaeological Review* 9-10, 15-21.

Byrne, R. and Horn, S.P. 1989. Prehistoric agriculture and forest clearance in the Sierra de los Tuxtlas, Veracruz, Mexico. *Palynology* 13, 181-93.

Cammas, C., Champagne, F., David, C., Desachy, B. and Guyard, L. 1995. Le probleme des «terres noires» sur les sites urbains tardo-antiques et médiévaux: réflexions et propositions méthodologiques à partir de l'exemple des fouilles du Collège de France à Paris. *Les Nouvelles de l'Archèologie* 61, 22-29.

Cammas, C., David, C. and Guyard, L. 1996. La question des Terres Noires dans les sites tardo-antiques et médievaux: le cas du College de France (Paris, France). In Broglio, G.P., Gelichi, S., Francovich, R., Hodges, R. and Steuer, H. (eds.) *Archaeology and History of the Middle Ages, Volume 14*, 89-93. Forlì: ICPPS.

Canti, M.G. 1997. An investigation of microscopic calcareous spherulites from herbivore dungs. *Journal of Archaeological Science* 24, 219-31.

Canti, M.G. 1998. The micromorphological identification of faecal spherulites from archaeological and modern materials. *Journal of Archaeological Science* 25, 435-44.

Carter, S.P. 1993-4. Radiocarbon dating evidence for the age of narrow cultivation ridges in Scotland. *Tools and Tillage* VII(2-3), 83-91.

Carter, S.P. and Davidson, D.A. 1998. An evaluation of the contribution of soil micromorphology to the study of ancient arable agriculture. *Geoarchaeology* 13(6), 535-47.

Carter, S.P., Tipping, R., Davidson, D., Long, D. and Tyler, A. 1997. A multiproxy approach to the function of postmedieval ridge-and-furrow cultivation in upland northern Britain. *The Holocene* 7(4), 447-56.

Catt, J.A. 1991. Soils as indicators of Quaternary climatic change in mid-latitude regions. *Geoderma* 51, 167-87.

Caulfield, S. 1978. Neolithic fields: the Irish evidence. In Bowen, H.C. and Fowler, P.J. (eds.) *Early Land Allotment*, 137-43. British Archaeological Reports,

British Series 48. Oxford: British Archaeological Reports.

Chadburn, A. 1987. The excavation: site 6 (the Dragon-fly ponds) 1984-5. In Cunliff, B. (ed.) *Hengistbury Head, Dorset. Volume 1: The Prehistoric and Roman Settlement, 3500 BC-AD 500*, 61-66 and 128-35. Oxford University Committee for Archaeology Monograph No. 13. Oxford: Oxford University Committee for Archaeology.

Chadburn, A. and Gardiner, J. 1985. A Grooved Ware pit and prehistoric spade marks on Hengistbury Head (Site 6), Dorset, 1984. *Proceedings of the Prehistoric Society* 51, 315-18.

Chadwick, A.M. 1997. Towards a social archaeology of later prehistoric and Romano-British field systems in South Yorkshire, West Yorkshire and Nottinghamshire. *http://www.assemblage.group.shef.ac.uk/2/2chad.html* (Access date: 18/08/11).

Chartres, C.J. and Mücher, H.J. 1989. The effect of fire on the surface properties and seed germination in two shallow monoliths from a rangeland soil subjected to simulated raindrop impact and water erosion. *Earth Surface Processes and Landforms* 14(5), 407-17.

Chen, Y., Tarchitzky, J., Brouwer, J., Morin, J. and Banin, A. 1980. Scanning electron microscope observations on soil crusts and their formation. *Soil Science* 130, 49-55.

Childe, V.G. 1929. *The Danube in Prehistory*. Oxford: Clarendon Press.

Childe, V.G. 1949. *Prehistoric Communities of the British Isles*. 3rd edition. London: W. & R. Chambers Ltd.

Clark, J.G.D. 1952. *Prehistoric Europe – the Economic Basis*. London: Methuen and Co. Ltd.

Clark, J.S. 1988a. Stratigraphic charcoal analysis on petrographic thin sections: application to fire history in northwestern Minnesota. *Quaternary Research* 30, 81-91.

Clark, J.S. 1988b. Particle motion and the theory of stratigraphic charcoal analysis: source area, transport, deposition, and sampling. *Quaternary Research* 30, 67-80.

Clemmensen, L.B., Murray, A.S., Bech, J.-H. and Clausen, A. 2001. Large-scale aeolian sand movement on the west coast of Jutland, Denmark in late Subboreal to early Subatlantic time – a record of climate change or cultural impact? *GFF* 123, 193-203.

Coles, J.M. 1973. *Archaeology by Experiment*. London: Hutchinson University Library.

Coles, J.M. 2010. *Experimental Archaeology*. Caldwell: Blackburn Press.

Collins, J.F. and Larney, F.J. 1987. Micromorphological observations of compacted horizons (cultivation pans) from various horizons in Irish tillage soils. In Federoff, N., Bresson, L.M. and Courty, M.A. (eds.) *Soil Micromorphology. Proceedings of the 7th International Working Meeting on Soil Micromorphology, Paris - July 1985,* 451-58. Plaisir: Association Française pour l'etude du Sol.

Collis, J. 1978. Fields and settlements on Shaugh Moor, Dartmoor. In Bowen, H.C. and Fowler, P.J. (eds.) *Early Land Allotment*, 23-27. British Archaeological Reports, British Series 48. Oxford: British Archaeological Reports.

Conry, M.J. 1971. Irish plaggen soils – their distribution, origin and properties. *Journal of Soil Science* 22(4), 401-16.

Conry, M.J. 1974. Plaggen soils, a review of man-made raised soils. *Soils and Fertilizers* 37, 319-26.

Conry, M.J. and Mitchell, G.F. 1971. The age of Irish plaggen soils. In Yaalon, D.H. (ed.) *Paleopedology: Origins, Nature and Dating of Paleosols*, 129-37. Jerusalem: ISSS and Israel Universities Press.

Conry, M.J., De Coninck, F. and Stoops, G. 1996. The properties, genesis and significance of a man-made iron pan podzol near Castletownbere, Ireland. *European Journal of Soil Science* 47, 279-84.

Cooney, G. 1997. Images of settlement and the landscape in the Neolithic. In Topping, P. (ed.) *Neolithic Landscapes*, 23-31. Neolithic Studies Group Seminar Papers 2. Oxbow Monograph 86. Oxford: Oxbow.

Coppock, J.T. 1964. *An Agricultural Atlas of England and Wales*. London: Faber and Faber Ltd.

Cornwall, I.W. 1953. Soil science and archaeology with illustrations from some British Bronze Age monuments. *Proceedings of the Prehistoric Society* 2(9), 129-147.

Cornwall, I.W. 1958. *Soils for the Archaeologist*. London: Phoenix House.

Cornwall, I.W. 1963. Appendix I: Soil samples from the Willerby Wold long barrow. In Manby, T.G. 'The excavation of the Willerby Wold long barrow, East Riding, Yorkshire'. *Proceedings of the Prehistoric Society* 29, 200-201.

Courty, M.A. 1992. Soil micromorphology in archaeology. In Pollard, A.M. (ed.) 'New developments in archaeological science'. *Proceedings of the British Academy* 77, 39-59. Oxford: Oxford University Press.

Courty, M.A. 1996. Application of soil micromorphology in archaeology. In Kooistra, M.J. (ed.) *European Training Course on Soil Micromorphology. II. Applications to Soil Micromorphology*. Wageningen: The Winand Staring Centre for Integrated Land, Soil and Water Research.

Courty, M.A. and Federoff, N. 1982. Micromorphology of a Holocene dwelling. *PACT* 7(II), 257-77.

Courty, M.A., Fedoroff, N. and Guilloré, P. 1987. Micromorphologie des sédiments archéologiques. In Miskovsky, J.C. and Curien, H. (eds.) *Géologie de la Préhistoire: Methodes, Techniques, Applications*, 439-77. Paris: Association pour l'Étude de l'Environnement Géologique de la Préhistoire.

Courty, M.A., Goldberg, P. and Macphail, R.I. 1989. *Soils and Micromorphology in Archaeology*. Cambridge: Cambridge University Press.

Courty, M.A., Goldberg, P. and Macphail, R.I. 1996. Ancient people – lifestyles and cultural patterns. In Wilding, L and Oleshko, K. (eds.) *Transactions of the 15th World Congress of Soil Science, Acapulco, Mexico, July 10-16, 1994*, 250-69. Acapulco: International Society of Soil Science.

Courty, M.A., Macphail, R.I. and Wattez, J. 1991. Soil micromorphological indicators of pastoralism; with special reference to Arene Candide, Finale Ligure, Italy. *Rivista di Studi Liguri* A. LVII(1-4), 127-50.

Crowther, J., Macphail, R.I. and Cruise, G.M. 1996. Short-term post-burial change in a humic rendzina soil, Overton Down Experimental Earthwork, Wiltshire, England. *Geoarchaeology* 11, 95-117.

Crummy, P., Shimmin, D., Carter, G. and Smith, N.A. 1992. Chapter 3: Excavations at Culver Street 1981-2 and 1984-5. In Crummy, P. *Excavations at Culver Street, the Gilberd School, and Other Sites in Colchester 1971-85*, 21-126. Colchester Archaeological Report 6. Colchester: Colchester Archaeological Trust Ltd. and English Heritage.

Cummins, T. 2007. *The Archaeology of Anthropogenic Soils in Ireland.* Unpublished PhD thesis. University College Dublin.

Currie, D.R. 1994. Micromorphology of a Native American cornfield. *Archaeology of Eastern North America* 22, 63-72.

Curwen, E. and Curwen, E.C. 1923. Sussex lynchets and their associated fieldways. *Sussex Archaeological Collections* 64, 1-65.

Curwen, E.C. and Hatt, G. 1953. *Plough and Pasture: the Early History of Farming*. New York: Henry Schuman Inc.

Cwynar, L.C. 1978. Recent history of fire and vegetation from annually laminated sediment of Greenleaf Lake, Algonquin Park, Ontario. *Canadian Journal of Botany* 56, 10-21.

Dalrymple, J.B. 1958. The application of soil micromorphology to fossil soils and other deposits from archaeological sites. *Journal of Soil Science* 9(2), 199-209.

Darmody, R.G. and Norton, L.D. 1994. Structural degradation of a prairie soil from long-term management. In Ringrose-Voase, A.J. and Humphreys, G.S. (eds.) *Soil Micromorphology: Studies in Management and Genesis. Proceedings of the 9th International Working Meeting on Soil Micromorphology, Townsville, Australia, July 1992*, 641-9. Developments in Soil Science 22. Amsterdam: Elsevier.

Datoo, B.A. 1978. Towards a reformulation of Boserup's theory of agricultural change. *Economic Geography* 54, 135-44.

Davidson, D.A. 1980. Erosion in Greece during the first and second millennia BC. In Cullingford, R.A., Davidson, D.A. and Lewin, J. (eds.) *Timescales in Geomorphology*, 143-58. Chichester: John Wiley and Sons.

Davidson, D.A. and Carter, S.P. 1998. Micromorphological evidence of past agricultural practices in cultivated soils: the impact of a traditional agricultural system on soils in Papa Stour, Shetland. *Journal of Archaeological Science* 25, 827-38.

Davidson, D.A. and Simpson, I.A. 1984. The formation of deep topsoils in Orkney. *Earth Surface Processes and Landforms* 9, 75-81.

Davidson, D.A., Carter, S.P. and Quine, T.A. 1992. An evaluation of micromorphology as an aid to archaeological interpretation. *Geoarchaeology* 7(1), 55-65.

Dawood, V. and Fitzpatrick, E.A. 1993. Some population sizes and effects of the Enchytraeidae (Oligochaeta) on soil structure in a selection of Scottish soils. *Geoderma* 56, 173-78.

de Olmedo Pujol, J.L. 1983. Anthropic influence on a Mediterranean brown soil. In Bullock, P. and Murphy, C.P. (eds.) *Soil Micromorphology, Volume 2: Soil Genesis. Proceedings of the 6th International Working Meeting on Soil Micromorphology, London 1981,* 583-88. Berkhamsted: AB Academic Publishers.

Dearing, J.A., Alström, K., Bergman, A., Regnell, J. and Sandgren, P. 1990. Recent and long-term records of soil erosion from southern Sweden. In Boardman, J., Foster, I.D.L. and Dearing, J.A. (eds.) *Soil Erosion on Agricultural Land*, 173-91. Chichester: John Wiley and Sons.

Denison, S. 1998. Prehistoric boundaries 'in medieval open fields'. *British Archaeology* 39, 4.

Desbiolles, J. 1993. *Development of semi-empirical models for tillage implements.* Unpublished PhD thesis. Silsoe College, Cranfield University.

Dewilde, B. 1983. *Twintig Eeuwan Vlas in Vlaanderen.* Tielft: Lannoo.

Dexter, A.R. 1976. Internal structure of tilled soil. *Journal of Soil Science* 27, 267-78.

Dexter, A.R. 1978. Tunnelling in soil by earthworms. *Soil Biology and Biochemistry* 10, 447-49.

Dexter, A.R. 1979. Prediction of soil structures produced by tillage. *Journal of Terramechanics* 16(3), 117-27.

Dexter, A.R. 1985. Shapes of aggregates from tilled layers of some Dutch and Australian soils. *Geoderma* 35(2), 91-107.

Diehl, W.J., Dabney, S.M., Romkens, M.J., Williams, D.L. and Hrantiz, J.M. 1995. Differences in earthworm populations among tillage systems in upland loess soils of northern Mississippi. Summary presented on the Agricultural Research Service TEKTRAN page at http://www.nalusda.gov/ttic/tektran/data/000006/33/0000063384.html.

Dimbleby, G.W. 1960. Appendix III: Pollen analysis. In Christie, P.M. Crig-a-mennis: a Bronze Age barrow at Liskey, Perranzabuloe, Cornwall, 94-96. *Proceedings of the Prehistoric Society* 26, 76-97.

Dimbleby, G.W. 1961. Soil pollen analysis. *Journal of Soil Science* 12, 1-11.

Dimbleby, G.W. 1985. *The Palynology of Archaeological Sites.* London: Academic Press.

Dixon, P. 1994. Field systems, rig and other cultivation remains in Scotland: the field evidence. In Foster, S. and Smout, T.C. (eds.) *The History of Soils and Field Systems*, 26-52. Aberdeen: Scottish Cultural Press.

Dockrill, S.J. and Gator, J.A. 1992. Tofts Ness: exploration and interpretation in a prehistoric landscape. In Spoerry, P. (ed.) *Geoprospection in the Archaeological Landscape*, 25-31. Oxford: Oxbow Books.

Dockrill, S.J. and Simpson, I. 1994. The identification of prehistoric anthropogenic soils in the Northern Isles using an integrated sampling strategy. *Archaeological Prospection* 1, 75-92.

Dockrill, S.J., Bond, J.M., Milles, A., Simpson, I. and Ambers, J. 1994. Tofts Ness, Sanday, Orkney. An integrated study of a buried Orcadian landscape. In Luff, R. and Rowley-Conwy, P. (eds.) *Whither Environmental Archaeology?*, 115-32. Oxbow Monograph 38. Oxford: Oxbow Books.

Dormaar, J.F., Pittman, U.J. and Spratt, E.D. 1979. Burning crop residues: effect on selected soil characteristics and long-term wheat yields. *Canadian Journal of Soil Science* 59(2), 79-86.

Dosedla, H.C. 1984. František Šach's contribution towards research on pre-industrial tilling implements in Austria. *Tools and Tillage* 5(1), 43-57.

Drewett, P. 1978. Field systems and land allotment in Sussex, 3rd millennium B.C. to 4th century A.D. In Bowen, H.C. and Fowler, P.J. (eds.) *Early Land Allotment*, 67-80. British Archaeological Reports, British Series 48. Oxford: British Archaeological Reports.

Drummond-Murray, J. 1994. *Wolseley Street, land adjacent to the fire station, SE1, London Borough of Southwark: An archaeological evaluation.* Museum of London Archaeology Service Report. London: Museum of London Archaeology Service.

Duchaufour, P. 1982. *Pedology* (English edition, translated by T. R. Paton). London: George Allen and Unwin.

Duncan, J.S. 1993. Landscapes of the self/landscapes of the other(s): cultural geography 1991-92. *Progress in Human Geography* 17(3), 367-77.

Edwards, K.J. 1982. Man, space and the woodland edge: speculations on the detection and interpretation of human impact in pollen profiles. In Bell, M. and Limbrey, S. (eds.) *Archaeological Aspects of Woodland Ecology*, 5-22. British Archaeological Reports, International Series 146. Oxford: British Archaeological Reports.

Edwards, K.J. 1983. Phosphate analysis of soils associated with the Old Kinord field and settlement system, Muir of Dinnet, Aberdeenshire. *Proceedings of the Society of Antiquaries of Scotland* 113, 620-27.

Edwards, W.M., Norton, L.D. and Redmond, C.E. 1988. Characterizing macropores that affect infiltration into non-tilled soil. *Soil Science Society of America Journal* 52, 483-87.

Eidt, R.C. 1984. *Advances in Abandoned Settlement Analysis: Application to Prehistoric Anthrosols in*

Colombia, South America. Milwaukee: The Center for Latin America, University of Wisconsin.

Ellis, S. and Mellor, A. 1995. *Soils and Environment*. London and New York: Routledge.

Englemark, R. 1992. A review of the farming economy in south Scania based on botanical evidence. In Larsson, L., Callmer, J. and Stjärnquist, B. (eds.) The archaeology of the cultural landscape. Field work and research in a South Swedish rural region. *Acta Archaeological Lundensia* Series No. 4°, 19, 369-75.

Englemark, R. and Linderholm, J. 1996. Prehistoric land management and cultivation. A soil chemical study. *Proceedings of the 6th Nordic Conference on the Application of Scientific Methods in Archaeology, Esbjerg 1993*, 315-22. Arkæologiske Rapporter fra Esbjerg Museum, No. 1. Esbjerg: Esbjerg Museum.

Entwistle, J.A., Abrahams, P.W. and Dodgshon, R.A. 1998. Multi-element analysis of soils from Scottish historical sites. Interpreting land-use history through the physical and geochemical analysis of soil. *Journal of Archaeological Science* 25, 53-68.

Entwistle, R. 1994. The environmental setting of the linear ditches system. In Bradley, R., Entwistle, R. and Raymond, F. *Prehistoric Land Divisions on Salisbury Plain*, 101-19. English Heritage Archaeological Report No. 2. London: English Heritage.

European Commission. 1995-2010. Glossary of Soil Terms. Joint Research Centre for Environment and Sustainability. http://eusoils.jrc.ec.europa.eu/esdb_archive/Glossary/Soil_Terms.html (Accessed 14/08/11)

Evans, C. and Serjeantson, D. 1988. The backwater economy of a fen-edge community in the Iron Age: the Upper Delphs, Haddenham. *Antiquity* 62(235), 360-70.

Evans, D.D. and Buol, S.W. 1968. Micromorphological study of soil crusts. *Soil Science Society of America Proceedings* 32, 19-22.

Evans, E.E. 1970. Introduction. In Gailey, A. and Fenton, A. (eds.) *The Spade in Northern and Atlantic Europe*, 1-9. Holywood and Belfast: Ulster Folk Museum and the Institute of Irish Studies, Queen's University Belfast.

Evans, J.G. 1971a. Notes on the environment of early farming communities in Britain. In Simpson, D.D.A. (ed.) *Economy and Settlement in Neolithic and Early Bronze Age Britain and Europe*, 11-26. Leicester: Leicester University Press.

Evans, J.G. 1971b. Habitat change on the calcareous soils of Britain: the impact of Neolithic man. In Simpson, D.D.A. (ed.) *Economy and Settlement in Neolithic and Early Bronze Age Britain and Europe*, 27-73. Leicester: Leicester University Press.

Evans, J.G., Atkinson, R.J.C., O'Connor, T. and Green, H.S. 1984. Stonehenge – the environment in the Late Neolithic and Early Bronze Age *and* a Beaker-age burial. *Wiltshire Archaeological and Natural History Magazine* 78, 7-30.

Evans, R. 1980. Mechanics of water erosion and their spatial and temporal controls: an empirical viewpoint. In Kirkby, M.J. and Morgan, R.P.C. (eds.) *Soil Erosion*, 109-28. Chichester: J. Wiley.

Evershed, R.P., Bethell, P.H., Reynolds, P.J. and Walsh, N.J. 1997. 5β-Signmastanol and related 5β-stanols as biomarkers of manuring: analysis of modern experimental material and assessment of the archaeological potential. *Journal of Archaeological Science* 24, 485-95.

Everton, A. and Fowler, P. 1978. Pre-Roman ard-marks at Lodge Farm, Alveston, Avon: a method of analysis. In Bowen, H.C. and Fowler, P.J. (eds.) *Early Land Allotment*, 179-85. British Archaeological Reports, British Series 48. Oxford: British Archaeological Reports.

Exaltus, R.P. and Miedema, R. 1994. A micromorphological study of four Neolithic sites in the Dutch coastal provinces. *Journal of Archaeological Science* 21(3), 289-301.

Fasham, P.J. 1980. Excavations on Bridget's and Burntwood Farms, Itchen Valley Parish, Hampshire, 1974. MARC 3 Sites R5 and R6. *Proceedings of the Hampshire Field Club Archaeological Society* 36, 37-86.

Fasham, P. and Hanworth, R. 1978. Ploughmarks, Roman roads and motorways. In Bowen, H.C. and Fowler, P. (eds.) *Early Land Allotment*, 175-77. British Archaeological Reports, British Series 48. Oxford: British Archaeological Reports.

Feachem, R.W. 1973. Ancient agriculture in the highland zone of Britain. *Proceedings of the Prehistoric Society* 39, 332-53.

Fechner, K. 1996. Paléoenvironnement et vie quotidienne. In Remy, H. and Soumoy, M. (eds.) *Sur la Voie de l'Histoire Archéologie et TGV*, 127-28. Etudes et Documents, séries Fouilles, 2. Namur: Ministère de la Région wallonne, Direction générale de l'Aménagement du Territioire, du Logement et du Patrimoine, Division du Patrimoine.

Fechner, K. 2000. Comparaison des histoires pédo-sédimentaires des fossés protohistoriques étudiés en Belgique et dans le Nord de la France. *Revue archéologique de Picardie* 1(2), 23-56.

Fechner, K., Louwagie, G. and Langohr, R. 1998. Données nouvelles sur l'agriculture protohistorique le long du tracé occidental du TGV en Wallonie. Les sites de Chièvres (Ht), d'Arbre (Ht) et quelques sites de comparaison. *Lunula. Archaeologica Protohistorica* 6, 77-82.

Fenton, A. 1994. Field systems and cultivating implements. In Foster, S. and Smout, T.C. (eds.) *The History of Soils and Field Systems*, 75-82. Aberdeen: Scottish Cultural Press.

Fisher, P.F. 1982. A review of lessivage and Neolithic cultivation in southern England. *Journal of Archaeological Science* 9(3), 299-304.

Fisher, P.F. and Macphail, R.I. 1985. Studies of archaeological soils and deposits by micromorphological techniques. In Fieller, N.R.J., Gilbertson, D.G. and Ralph, N.G.A. (eds.) *Palaeoenvironmental Investigations: Research Design, Methods and Data Analysis*, 93-125. Symposia of the Association for Environmental Archaeology No. 5A. British Archaeological Reports, International Series 258. Oxford: British Archaeological Reports.

Fitzpatrick, E.A. 1971. *Pedology: A Systematic Approach to Soil Science.* Edinburgh: Oliver and Boyd.

Fitzpatrick, E.A. 1984. *Micromorphology of Soils.* London: Chapman and Hall.

Fitzpatrick, E.A. 1986. *An Introduction to Soil Science.* London: Longmans.

Fitzpatrick, E.A. 1993. *Soil Microscopy and Micromorphology.* Chichester: John Wiley and Sons.

Fleming, A. 1978. The Dartmoor Reaves. In Bowen, H. C. and Fowler, P.J. (eds.) *Early Land Allotment*, 17-21. British Archaeological Reports, British Series 48. Oxford: British Archaeological Reports.

Fleming, A. 1985. Land tenure, productivity and field systems. In Barker, G. and Gamble, C. (eds.) *Beyond Domestication in Prehistoric Europe,* 129-46. London: Academic Press.

Fleming, A. 1988. *The Dartmoor Reaves: Investigating Prehistoric Land Divisions.* London: B.T. Batford Ltd.

Fleming, A. 1989a. Coaxial field systems in later British prehistory. In Nordström, H-Å. and Knape, A. (eds.) *Bronze Age Studies*, 151-62. Transactions of the British-Scandinavian Colloquium in Stockholm, May 10-11, 1985. The Museum of National Antiquities, Stockholm, Studies 6. Stockholm: Statens Historiska Museum.

Fleming, A. 1989b. The genesis of coaxial field systems. In van der Leeuw, S.E. and Torrence, R. (eds.) *What's New? A Closer Look at the Process of Innovation*, 63-81. London: Unwin Hyman.

Fleming, A. 1995. St. Kilda: stone tools, dolerite quarries and long-term survival. *Antiquity* 69, 25-35.

Folk, R.L. 1959. Practical petrographic classification of limestones. *American Association of Petrological Geology Bulletin* 43, 1-38.

Fowler, P.J. 1967. The archaeology of Fyfield and Overton Downs, Wiltshire. (3rd interim report). *Wiltshire Archaeological and Natural History Magazine* 62, 16-33.

Fowler, P.J. 1971. Early prehistoric agriculture in Western Europe: some archaeological evidence. In Simpson, D.D.A. (ed.) *Economy and Settlement in Neolithic and Early Bronze Age Britain and Europe,* 153-82. Leicester: Leicester University Press.

Fowler, P.J. 1981a. Agriculture and rural settlement. In D.M. Wilson (ed.) *The Archaeology of Anglo-Saxon England,* 23-48. Cambridge: Cambridge University Press.

Fowler, P.J. 1981b. Wildscape to landscape: 'enclosure' in prehistoric Britain. In Mercer, R. *Farming Practice in British Prehistory,* 9-54. Edinburgh: Edinburgh University Press.

Fowler, P.J. 1983. *The Farming of Prehistoric Britain.* Cambridge: Cambridge University Press.

Fowler, P.J. 2000. *Landscape Plotted and Pieced: landscape history and local archaeology in Fyfield and Overton Down, Wiltshire.* Reports of the Research Committee of the Society of Antiquaries of London, No. 64. London: Society of Antiquaries.

Fowler, P.J. and Evans, J.G. 1967. Plough-marks, lynchets and early fields. *Antiquity* 41, 289-301.

Frazer, J.G. 1925. *The Golden Bough. Part V. Spirits of the Corn and of the Wild. Volumes I-II.* 3rd edition. London: MacMillan and Co.

French, C.A.I. 1985-6a. Soil micromorphological analysis of the buried soils and stream deposits at Etton Woodgate and Etton, Cambridgeshire: some interim results. *Fenland Research* 3, 40-43.

French, C.A.I. 1985-6b. Soil micromorphological analysis of early prehistoric buried soils in the North Level, Cambridgeshire: some interim results. *Fenland Research* 3, 43-45.

French, C.A.I. 1988. The southwest Fen dyke survey project. *Antiquity* 62(235), 343-48.

French, C.A.I. 1990. Neolithic soils, middens and alluvium in the lower Welland valley. *Oxford Journal of Archaeology* 9(3), 305-11.

French, C.A.I. 2003. *Geoarchaeology in Action.* London and New York: Routledge.

French, C.A.I. 2005. Analysis of the soil deposits at Barnhouse and Maeshowe. In Richards, C. (ed.) *Dwelling among the monuments: the Neolithic village of Barnhouse, Maeshowe passage grave and surrounding monuments at Stenness, Orkney*, 371-81. McDonald Institute Monographs. Cambridge: McDonald Institute

French, C. and Lewis, H. 2001. The Depot site – soil micromorphological analysis. In Pryor, F. *The Archaeology and Environment of the Flag Fen Basin*, 20-22 and 437-42. Swindon: English Heritage.

French, C.A.I. and Marsh, P. 1999. Welland Bank Quarry, Lincolnshire, 1997: the soil micromorphological analysis of the dark earth and buried soil. Unpublished report. Heckington: Archaeological Project Services.

French, C.A.I. and Pryor, F. 1992. Floodplain gravels: buried Neolithic and Bronze Age landscapes along the fen margins. In Fulford, M. and Nichols, E. (eds.) *Developing Landscapes of Lowland Britain: The Archaeology of the British Gravels: A Review*, 63-77. London: Society of Antiquaries of London.

French, C.A.I. and Pryor, F. 1993. *The south-west fen dyke survey project 1982-1986.* East Anglian Archaeology Monograph 59. Cambridge: Cambridgeshire Archaeological Committee.

French, C., Ballantyne, R., Carrado, A., Cyganowski, C., Dodwell, N., Evans, C., Gdaniec, K., GSB Prospection, Hanks, B., Hill, JD, Lewis, H., Miracle, P., Oswald, A., Pattison, P. and Shell, C. 2004. Evaluation survey and excavation at Wandlebury Ringwork, Cambridgeshire, 1994-7. *Proceedings of the Cambridge Antiquarian Society* XCIII, 15-66.

French, C., Lewis, H., Allen, M.J., Green, M., Scaife, R. and Gardiner, J. 2007. *Prehistoric landscape development and human impact in the upper Allen valley, Cranborne Chase, Dorset.* McDonald Institute Monograph. Cambridge: McDonald Institute for Archaeological Research.

Fujiwara, H. 1993. Research into the history of rice cultivation using plant opal analysis in archaeology and paleoecology. In Pearsall, D.M. and Piperno, D. (eds.) *Current Research in Phytolith Analysis: Application*, 147-59. Philadelphia: MASCA, The University Museum of Archaeology and Anthropology, University of Pennsylvania.

Gaffney, V., Stancic, Z. and Watson, H. 1995. Moving from catchments to cognition: tentative steps towards a larger archaeological context for GIS. *Scottish Archaeological Review* 9/10, 41-64.

Gallagher, J.P. 1989. Agricultural intensification and ridged-field cultivation in the prehistoric upper Midwest of North America. In Harris, D.R. and Hillman, G.C. (eds.) *Foraging and Farming: The Evolution of Plant Exploitation*, 572-84. One World Archaeology 13. London: Unwin Hyman.

Gebhardt, A. 1990. *Evolution du paléopaysage agricole dans le nord-ouèst de la France: apport de la micromorphologie.* Unpublished PhD thesis, Université de Rennes I.

Gebhardt, A. 1991a. Soil micromorphology. *Tsukuba Archaeological Studies, March 2nd, 1991*, 79-98. Tsukuba: Institute of History and Archaeology, University of Tsukuba.

Gebhardt, A. 1991b. Analyse micromorphologique des modifications apportées à la structure du sol par différents outils. *Archéologie Aujourd'hui, Archéologie Expérimental. Le terre* T2, 223-28.

Gebhardt, A. 1992. Micromorphological analysis of soil structural modifications caused by different cultivation implements. In *Préhistoire de l'Agriculture: Nouvelles Approches Expérimentales et Ethnographiques*, 373-81. Monographie du CRA No. 6. Paris: CNRS.

Gebhardt, A. 1993. Micromorphological evidence of soil deterioration since the mid-Holocene at archaeological sites in Brittany, France. *The Holocene* 3, 333-41.

Gebhardt, A. 1995. Soil micromorphological data from traditional and experimental agriculture. In Barham, A.J. and Macphail, R.I. (eds.) *Archaeological Sediments and Soils: Analysis, Interpretation and Management*, 25-39. London: Institute of Archaeology, University College London.

Gebhardt, A. 1996. Fossés parcellaires et micromorphologie des sols: une aide à l'interprétation archéologique d'un paysage agraire? In Castelletti, L. and Cremaschi, M. (eds.) *Paleoecology. XIII International Congress of Prehistoric and Protohistoric Sciences, Forlì (Italia), 8/14 September 1996. Colloquia: V. Numerical analysis and image processing in archaeobotany; VI. Micromorphology of deposits of anthropogenic origin*, 107-13. Forlì: A.B.A.C.O.

Gebhardt, A. 2007. Impact of charcoal making activities on a soil profile: the micromorphological point of view. *Archéosciences* 31, 127-37.

Geiss, J.W. 1973. Biogenic silica in selected species of deciduous angiosperms. *Soil Science* 116, 113-19.

Gerrard, J. 1992. *Soil Geomorphology: An Integration of Pedology and Geomorphology*. London: Chapman and Hall.

Gibbs, R.J. and Reid, J.B. 1988. A conceptual model of changes in soil structure under different cropping systems. In Stewart, B.A. (ed.) *Advances in Soil Science*, Volume 8, 123-49. New York: Springer-Verlag.

Gibson, D.B. 1988. Agro-pastoralism and regional social organisation in early Ireland. In Gibson, D.B. and Geselowitz, M.N. (eds.) *Tribe and Polity in Late Prehistoric Europe: Demography, Production and Exchange in the Evolution of Complex Social Systems*, 41-68. New York and London: Plenum Press.

Glob, P.V. 1945. Ploughs of the Døstrup type found in Denmark. *Acta Archaeologica* 16, 93-111.

Glob, P.V. 1951. *Ard og Plov i Nordens Oldtid* (in Danish with English summary). Jysk Arkæologisk selskabs Skrifter Bind 1. Aarhus: Universitetsforlaget i Aarhus.

Goldberg, P. 1980. Micromorphology in archaeology and prehistory. *Paleorient* 6, 159-64.

Goldberg, P. 1983. Applications of micromorphology in archaeology. In Bullock, P. and Murphy, C.P. (eds.) *Soil Micromorphology, Volume 1: Techniques and Applications. Proceedings of the 6th International Working Meeting on Soil Micromorphology, London 1981*, 139-50. Berkhamsted: AB Academic Publishers.

Goldberg, P. 1992. Micromorphology, soils, and archaeological sites. In Holliday, V.T. (ed.) *Soils in Archaeology: Landscape Evolution and Human Occupation,* 145-67. Washington and London: Smithsonian Institution Press.

Goldberg, P. and Macphail, R.I. 2006. *Practical and Theoretical Geoarchaeology*. Malden: Blackwell.

Goldberg, P., Lev-Yadun, S. and Bar-Yosef, O. 1994. Petrographic thin sections of archaeological sediments: a new method for palaeobotanical studies. *Geoarchaeology: An International Journal* 9(3), 243-57.

Goodman-Elgar, M. 2008. Evaluating soil resilience in long-term cultivation: a study of pre-Columbian terraces from the Paca Valley, Peru. *Journal of Archaeological Science* 35, 3072-86.

Goody, J. 1976. *Production and Reproduction. A Comparative Study of the Domestic Domain.* Cambridge: Cambridge University Press.

Goody, J. 1977. *The Domestication of the Savage Mind.* Cambridge: Cambridge University Press.

Goudie, A.S. 1981. *The Human Impact: Man's Role in Environmental Change.* Oxford: Basil Blackwell.

Govers, G., Quine, T.A., Desmet, P.J.J. and Walling, D.E. 1996. The relative contribution of soil tillage and overland flow erosion to soil redistribution on agricultural land. *Earth Surface Processes and Landforms* 21, 929-46.

Govers, G., Vandaele, K., Desmet, P., Poesen, J. and Bunte, K. 1994. The role of tillage in soil redistribution on hillslopes. *European Journal of Soil Science* 45, 469-78.

Greenland, D.J. 1977. Soil damage by intensive arable cultivation: temporary or permanent? *Philosophical Transactions of the Royal Society of London* B281, 193-208.

Grigg, D. 1979. Ester Boserup's theory of agrarian change: a critical review. *Progress in Human Geography* 3, 65-84.

Hall, D. 1981. The origins of open-field agriculture - the archaeological fieldwork evidence. In Rowley, T. (ed.) *The Origins of Open-field Agriculture*, 22-38. London: Croom Helm.

Hall, N.W. 1990. Micromorphology and complementary assessments of soil structure description and their relationships to the length of time under tillage and calcium carbonate contents. In Douglas, L. A. (ed.) *Soil Micromorphology: A Basic and Applied Science. Proceedings of the 8th International Working Meeting of Soil Micromorphology, San Antonio, Texas - July 1988*, 53-60. Amsterdam: Elsevier Science Publishers B.V.

Halliday, S.P., Hill, P.J. and Stevenson, J.B. 1981. Early agriculture in Scotland. In Mercer, R. (ed.) *Farming Practice in British Prehistory*, 55-65. Edinburgh: Edinburgh University Press.

Hansen, H.-O. 1969. Experimental ploughing with a Døstrup ard replica. *Tools and Tillage* 1(2), 67-92.

Hansen, H.-O. 1988. *Lejre Forsøgscenter*. Forsøg med fortiden 1. Lejre: Historisk-Arkæologisk Forsøgscenter.

Harding, J.M. 1964. Interim report on the excavation of a late Bronze Age homestead in Weston Wood, Albury, Surrey. *Surrey Archaeological Collections* 61, 10-17.

Harlan, J.R. 1995. *The Living Fields: Our Agricultural Heritage*. Cambridge: Cambridge University Press.

Harris, D.R. 1989. An evolutionary continuum of people-plant interaction. In Harris, D.R. and Hillman, G.C. (eds.) *Foraging and Farming: The Evolution of Plant Exploitation*, 11-26. London: Unwin Hyman.

Hassink, J., Bouwman, L.A., Zwart, K.B., Bloem J. and Brussaard, L. 1993. Relationships between soil texture, physical protection of organic matter, soil biota and C and N mineralization in grassland soils. *Geoderma* 57, 105-28.

Hatt, G. 1949. *Oldtidsagre*. Det kongelige danske Videnskabernes Selskab, Arkæologisk-kunsthistoriske Skrifter II, 1. København: I Kommission hos Munksgaard.

Havinga, A.J. 1974. Problems in the interpretation of pollen diagrams of mineral soils. *Geologie en Mijnbouw* 53, 449-53.

Havinga, A.J. 1984. A 20-year experimental investigation into the differential corrosion susceptibility of pollen and spores in various soil types. *Pollen et Spores* 26, 541-58.

Hey, G. 1997. Neolithic settlement at Yarnton, Oxfordshire. In Topping, P. (ed.) *Neolithic Landscapes*, 99-111. Neolithic Studies Group Seminar Papers 2. Oxbow Monograph 86. Oxford: Oxbow Books.

Hodder, I. 1990. *The Domestication of Europe: Structure and Contingency in Neolithic Societies*. Oxford: Basil Blackwell.

Holden, N.M. 1995. Temporal variation in ped shape in an old pasture soil. *Catena* 24, 1-11.

Holden, E.W. and Bradley, R.J. 1975. A late Neolithic site at Rackham. *Sussex Archaeological Collections* 113, 85-103.

Houyuan, L., Naiqui, W. and Baozhu, L. 1997. Recognition of rice phytoliths. In Pinilla, A., Juan-Tresserras, J. and Machado, M.J. (eds.) *Estado Actual de los Estudios de Fitolitos en Suelos y Plantas*, 159-74. Consejo Superior de Investigaciones Científicas Monografias 4. Madrid: Centro de Ciencias Medoambientales del Consjo Superior de Investigaciones Científicas.

Hunt, R.C. 2000. Labor productivity and agricultural development: Boserup revisited. *Human Ecology* 28(2):251-277.

Huttunen, P. 1980. Early land use, especially slash-and-burn cultivation in the commune of Lammi, southern Finland, interpreted mainly using pollen and charcoal analyses. *Acta Botanica Fennica* 113, 1-45.

Imeson, A.C. and Vis, M. 1984. The output of sediments and solutes from forested and cultivated clayey drainage basins in Luxembourg. *Earth Surface Processes and Landforms* 9, 585-94.

Imeson, A.C., Kwaad, F.J.P.M. and Mücher, H.J. 1980. Hillslope processes and deposits in forested areas of Luxembourg. In Cullingford, R.A., Davidson, D.A. and Lewin, J. (eds.) *Timescales in Geomorphology*, 31-42. Chichester: John Wiley and Sons.

Isaac, E. 1970. *Geography of Domestication*. Englewood Cliffs: Prentice-Hall Inc.

Iversen, J. 1941. Land occupation in Denmark's Stone Age. *Danmarks Geologiske Undersøgelse II* 66, 1-68.

Iversen, J. 1949. The influence of prehistoric man on vegetation. *Danmarks Geologiske Undersøgelse IV* 6, 1-25.

Jackson, J.W. 1957. Derbyshire lynchets. *Journal of the Derbyshire Archaeological and Natural History Society* 77, 62-63.

Jenny, H. 1941. *Factors of Soil Formation*. New York: McGraw-Hill.

Jewell, P.A. and Dimbleby, G.W. (eds.) 1966. The experimental earthwork on Overton Down, Wiltshire, England: the first four years. *Proceedings of the Prehistoric Society* 32, 313-42.

Jobey, G. 1968. Excavations of cairns at Chatton Sandyford, Northumberland. *Archæologica Æliana* 46, 4th series, 5-50.

Johnson, D.L. and Hansen, K.L. 1974. The effects of frost-heaving on objects in soils. *Plains Anthropologist* 19(64), 81-98.

Johnston, R. 2005. Pattern without a plan: rethinking the Bronze Age coaxial field systems on Dartmoor, south-west England. *Oxford Journal of Archaeology* 24(1), 1-21.

Jones, M. 1986. *England Before Domesday*. London: B.T. Batsford Ltd.

Jones, M. 1988. The arable field: a botanical battleground. In M. Jones (ed.) *Archaeology and the Flora of the British Isles*, 86-91. Oxford: Oxford University Committee for Archaeology.

Jones, M. 1989. Agriculture in Roman Britain: the dynamics of change. In M. Todd (ed.) *Research on Roman Britain, 1960-89*, 127-34. Britannian Monograph Series No. 11. London: The Society for the Promotion of Roman Studies.

Jones, M. 1992. 3. Food production and consumption. In Jones, R.F.J. (ed.) *Roman Britain: Recent Trends*, 21-27. Sheffield: J.R. Collis Publications, Department of Archaeology and Prehistory, University of Sheffield.

Jones, S. 1997. *The Archaeology of Ethnicity: Constructing Identities in the Past and Present.* London: Routledge.

Jongerius, A. 1970. Some morphological aspects of regrouping phenomena in Dutch soils. *Geoderma* 4, 311-31.

Jongerius, A. 1983. The role of micromorphology in agricultural research. In Bullock, P. and Murphy, C.P. (eds.) *Soil Micromorphology, Volume 1: Techniques and Application. Proceedings of the 6th International Working Meeting on Soil Micromorphology, London 1981*, 111-38. Berkhamsted: AB Academic Publishers.

Jongerius, A. and Jager, A. 1964. The morphology of humic gley soils (Orthic Haplaquolls) under different land use. In Jongerius, A. (ed.) *Soil Micromorphology. Proceedings of the 2nd International Working Meeting on Soil Micromorphology, Arnhem, the Netherlands, September 22-25, 1964*, 491-503. Amsterdam: Elsevier.

Kandeler, E. and Murer, E. 1993. Aggregate stability and soil microbial processes in a soil with different cultivation. *Geoderma* 56, 503-13.

Keith-Lucas, M. 1986. Neolithic impact on vegetation and subsequent vegetational development at Scord of Brouster. In Whittle, A., Keith-Lucas, M., Milles, A., Noddle, B., Rees, S. and Romans, J.C.C. *Scord of Brouster. An Early Agricultural Settlement on Shetland. Excavations 1977-1979*, 92-118. Oxford University Committee for Archaeology Monograph No. 9. Oxford: Oxford University Committee for Archaeology.

Kemp, R.A. 1985. *Soil Micromorphology and the Quaternary.* Quaternary Research Association Technical Guide No. 2. London: Birkbeck College and Quaternary Research Association.

Kemp, R.A., Whiteman, C.A. and Rose, J. 1993. Palaeoenvironmental and stratigraphic significance of the Valley Farm and Barnham soils in eastern England. *Quaternary Science Reviews* 12, 833-48.

Kooistra, M.J. 1987. The effects of compaction and deep tillage on soil structure in a Dutch sandy loam soil. In Federoff, N., Bresson, L.M. and Courty, M.A. (eds.) *Soil Micromorphology. Proceedings of the 7th International Working Meeting on Soil Micromorphology, Paris – July 1985*, 445-50. Plaisir: L'Association Français pour l'Étude du Sol.

Kooistra, M.J., Bouma, J., Boersma, O.H. and Jager, A. 1984. Physical and morphological characterization of undisturbed and disturbed ploughpans in a sandy loam soil. *Soil and Tillage Research* 4, 405-17.

Kooistra, M.J., Bouma, J., Boersma, O.H. and Jager, A. 1985. Soil-structure differences and associated physical properties of some loamy Typic Fluvaquents in the Netherlands. *Geoderma* 36(3-4), 215-28.

Kooistra, M.J., Juo, A.S.R. and Schoonderbeek, D. 1990. Soil degradation in cultivated alfisols under different management systems in southwestern Nigeria. In Douglas, L.A. (ed.) *Soil Micromorphology: A Basic and Applied Science. Proceedings of the 8th International Working Meeting of Soil Micromorphology, San Antonio, Texas - July 1988*, 61-9. Developments in Soil Science 19. Amsterdam: Elsevier Science Publishers B. V.

Kooistra, M.J., Tovey, N.K. and Pagliai, M. 1996. Agricultural applications. In Kooistra, M.J. (ed.) *European Training Course on Soil Micromorphology. II. Applications to Soil Micromorphology*, 1-5. Wageningen: Winand Staring Centre for Integrated Land, Soil and Water Research.

Korhonen, T. 1985. The effects of agricultural technology on the image and structure of the rural landscape in Finland. *Acta Museorum Agriculturae Pragae* 18(1), 138-149.

Kubiena, W.L. 1938. *Micropedology*. Ames: Collegiate Press.

Kwaad, F.J.P.M. and Mücher, H.J. 1979. The formation and evolution of colluvium on arable land in northern Luxembourg. *Geoderma* 22(3), 173-92.

Lagerås, P. and Sandgren, P. 1994. The use of mineral magnetic analyses in identifying middle and late Holocene agriculture – a study of peat profiles in Småland, southern Sweden. *Journal of Archaeological Science* 21, 687-97.

Lamb, R.C. and Rees, S.E. 1981. Ard cultivation at Sumburgh, Shetland. *Tools and Tillage* 4(2), 117-21.

Langohr, R. 1993. Types of tree windthrow, their impact on the environment and their importance for the understanding of archaeological excavation data. *Helenium* 33(1), 36-49.

Le Bissonnais, Y. 1996. Aggregate stability and assessment of soil crustability and erodibility: I. theory and methodology. *European Journal of Soil Science* 47, 425-37.

Lee, H. 2009. Soil micromorphology and the recognition of ancient wet rice farming: a case study from Gulhwa, South Korea. In Thiemeyer, H. (ed.) *Archaeological Soil Micromorphology: Contributions to the Archaeological Soil Micromorphology Working Group Meeting, 3rd to 5th April 2008*, 59-70. Frankfurter geowissenschaftliche Arbeiten. Serie D, Physische Geographie, Volume 30. Frankfurt: Institut für Physische Geographie, Johann Wolfgang Goethe-Universitat.

Lee, H. 2011. *The Agricultural Land Use Dynamics Associated with the Advent of Paddy Rice Cultivation in Bronze Age South Korea*. Unpublished PhD thesis. University of Cambridge.

Lee, K.E. 1985. *Earthworms: Their Ecology and Relationships with Soils and Land Use*. New York: Academic Press.

Leica Heerbrugg AG. 1991. *Wild MPS46/52 User Manual*. Heerbrugg: Wild Heerbrugg.

Lerche, G. 1977. Double paddle-spades in prehistoric contexts in Denmark. *Tools and Tillage* 3(2), 111-24.

Lerche, G. 1982. Additional comments on the Lindholm Høje field. *Tools and Tillage* 4(2), 185-91.

Lerche, G. 1984. Ridged fields and profiles of plough-furrows. Ploughing practices in medieval and post-medieval times. A study in experimental archaeology. *Tools and Tillage* 5(3), 131-56.

Lerche, G. 1994. *Ploughing Implements and Tillage Practices in Denmark from the Viking Period to about 1800 Experimentally Substantiated.* The Royal Danish Academy of Sciences and Letters' Commission for Research on the History of Agricultural Implements and Field Structures Publication No. 8. Herning: Poul Kristensen.

Lerche, G. and Steensberg, A. 1980. *Agricultural Tools and Field Shapes.* International Secretariat for Research on the History of Agricultural Implements Publication No. 3. Copenhagen: National Museum of Denmark.

Leser, P. 1931. Entstehung und Verbreitung des Pfluges. *Anthropos* B.111, 3.

Lewis, H.A. 1998. *The Characterisation and Interpretation of Ancient Tillage Practices through Soil Micromorphology: A Methodological Study.* Unpublished PhD thesis, University of Cambridge.

Lewis, H.A. 2002. An investigation of ancient cultivation remains at Hengistbury Head Site 6, Christchurch, Dorset. *Proceedings of the Prehistoric Society* 68, 83-102.

Lewis, H. 2003. Soil micromorphological study of a spade mark in Pit 77. In French, C. *Wandlebury Ringwork, Cambridgeshire, 1994-7: Evaluation, Survey and Excavation*, 51-56. Cambridge Archaeological Unit Report 553. Report for Cambridgeshire County Council, English Heritage and Cambridge Preservation Trust.

Lewis, H. 2008. 'The pleasant land of counterpane': linking site-specific archaeological land use to the landscape of prehistoric field systems. In Chadwick, A.M. (ed.) *Recent Approaches to the Archaeology of Land Allotment*, 239-50. British Archaeological Reports 1875. Oxford: BAR Publishing.

Lewis, H. 2010. Connotations of land use in landscape archaeology. In Lewis, H. and Semple, S. (eds.) *Perspectives in Landscape Archaeology. Papers presented at Oxford 2003-2005*, 50-82. British Archaeological Reports, International Series 2103. Oxford: BAR Publishing.

Lewis, H. and French. C.A.I. 2006. The soil micromorphological analyses. In Cromarty, A.M., Barclay, A. Lambrick, G. and Robinson, M. *Late Bronze Age Ritual and Habitation on a Thames Eyot at Whitecross Farm, Wallingford: The Archaeology of the Wallingford Bypass 1986-92*, 194-96. Thames Valley Landscapes Monograph 22. Oxford: Oxford Archaeology.

Limbrey, S. 1975. *Soil Science and Archaeology.* London: Academic Press.

Limbrey, S. 1992. Micromorphological studies of buried soils and alluvial deposits in a Wiltshire river valley. In Needham, S. and Macklin, M.G. (eds.) *Alluvial Archaeology in Britain*, 53-64. Oxbow Monograph 27. Oxford: Oxbow Books.

Linderholm, J. and Lundberg, E. 1994. Chemical characterisation of various archaeological soil samples using main and trace elements determined by Inductively Coupled Plasma Atomic Emission Spectrometry. *Journal of Archaeological Science* 21, 303-14.

Lindquist, S.-O. 1974. The development of the agrarian landscape on Gotland during the early Iron Age. *Norwegian Archaeological Review* 7(1), 6-32.

Lindqvist, C. 1987. Analogies in the spread of Neolithic and later innovations over Europe and Africa. In Burenhult, G., Carlsson, A., Hyenstrand, A. and Sjøvold, T. (eds.) *Theoretical Approaches to Artefacts, Settlements and Society. Studies in Honour of Mats P. Malmer, Part i*, 177-97. British Archaeological Reports International Series 366(i). Oxford: British Archaeological Reports.

Liversage, D. and Robinson, D. 1992-3. Prehistoric settlement and landscape development in the Sandhill belt of southern Thy. *Journal of Danish Archaeology* 11, 39-56.

Liversage, D., Munro, M.A.R., Courty, M.A. and Nørnberg, P. 1987. Studies of a buried Iron Age field. *Acta Archaeologica* 56, 55-84.

Livingston, S.J., Norton, L.D. and West, L.T. 1990. Effect of long-term cultivation on aggregate stability, organic carbon distribution, and porosity of two soil series. In Douglas, L.A. (ed.) *Soil Micromorphology: A Basic and Applied Science. Proceedings of the 8th*

International Working Meeting of Soil Micromorphology, San Antonio, Texas - July 1988, 89-95. Developments in Soil Science 19. Amsterdam: Elsevier Science Publishers B.V.

Lüning, J. and Meurers-Balke, J. 1980. Experimenteller Getreideanbau im Hambacher Forst, Gemeinde Elsdorf, Kr. Bergheim/Rheinland. *Bonner Jahrbücher* 180, 306-44.

Lustig-Arecco, V. 1975. *Technology. Strategies for Survival*. New York: Holt, Rinehart and Winston Inc.

MacDonald, G.M., Larsen, C.P.S., Szeicz, J.M. and Moser, K. 1991. The reconstruction of boreal forest fire history from lake sediments: a comparison of charcoal, pollen, sedimentological and geochemical indices. *Quaternary Science Reviews* 10(1), 53-71.

Mackie-Dawson, L.A., Mullins, C.E., Goss, M.J., Court, M.N. and Fitzpatrick, E.A. 1989. Seasonal changes in the structure of clay soils in relation to soil management and crop type. II. Effects of cultivation and cropping at Compton Beauchamp. *Journal of Soil Science* 40, 283-92.

Macphail, R.I. 1981. Soil and botanical studies of the Dark Earth. In Jones, M. and Dimbleby, G.W. (eds.) *The Environment of Man: the Iron Age to the Anglo-Saxon Period*, 309-31. British Archaeological Reports, British Series 87. Oxford: British Archaeological Reports.

Macphail, R.I. 1983. The micromorphology of dark earth from Gloucester, London and Norwich: an analysis of urban anthropogenic deposits from the Late Roman to Early Medieval periods in England. In Bullock, P. and Murphy, C.P. (eds.) *Soil Micromorphology, Volume 1: Techniques and Applications. Proceedings of the 6th International Working Meeting on Soil Micromorphology, London 1981*, 245-52. Berkhamsted: AB Academic Publishers.

Macphail, R.I. 1986a. *Soil Report on Hazleton Long Cairn, Gloucestershire, Volumes 1 and 2*. Ancient Monuments Laboratory Report No. 4897. London: Historic Buildings and Monuments Commission for England.

Macphail, R.I. 1986b. Palaeosols in archaeology: their role in understanding Flandrian pedogenesis. In Wright, V.P. (ed.) *Palaeosols: Their Recognition and Interpretation*, 263-90. Oxford: Blackwell Scientific Publishers.

Macphail, R.I. 1987a. A review of soil science in archaeology in England. In Keeley, H.C.M. (ed.) *Environmental Archaeology: A Regional Review. Volume II*, 332-79. Historic Buildings and Monuments Commission for England Occasional Paper No. 1. London: Historic Buildings and Monuments Commission for England.

Macphail, R I. 1987b. *Soil Report on the Late Bronze Age Midden at Potterne, near Devises, Wiltshire*. Unpublished Ancient Monuments Laboratory Report No. 144/87. London: Historic Buildings and Monuments Commission for England.

Macphail, R.I. 1987c. *Soil Report on the Cairn and Field System at Chysauster, Penzance, Cornwall*. Ancient Monuments Laboratory Report No. 111/87. London: Historic Buildings and Monuments Commission for England.

Macphail, R.I. 1987d. The soil micromorphology of tree subsoil hollows. *Circaea* 5(1), 14-17.

Macphail, R.I. 1989. *Soil Report on Trethellan Farm, Newquay, Cornwall*. Ancient Monuments Laboratory Report No. 69/89. London: Historic Buildings and Monuments Commission for England.

Macphail, R.I. 1990a. Soil micromorphological evidence of the impact of ancient agriculture. *14th International Congress of Soil Science Transactions, Volume 7*, 264-69. Kyoto: International Society of Soil Science.

Macphail, R.I. 1990b. Soil history and micromorphology. In Bell, M. *Brean Down Excavations 1983-1987*, 187-96, Microfiche. English Heritage Archaeological Report No. 15. London: Historic Buildings and Monuments Commission for England.

Macphail, R.I. 1990c. The soils. In Saville, A. *Hazleton North, Gloucestershire, 1979-82*, 223-26, Appendices 14-16. English Heritage Archaeological Report No. 13. London: Historic Buildings and Monuments Commission for England.

Macphail, R.I. 1992a. Soil micromorphological evidence of ancient soil erosion. In Bell, M. and Boardman, J. (eds.) *Past and Present Soil Erosion: Archaeological and Geological Perspectives*, 197-215. Oxbow Monograph 22. Oxford: Oxbow Books.

Macphail, R.I. 1992b. Soil studies. In Crummy, P. *Excavations at Culver Street, the Gilberd School, and other sites in Colchester 1971-85*, 273-5, 703-8. Colchester Archaeological Report 6. Colchester: Colchester Archaeological Trust Ltd. and English Heritage.

Macphail, R.I. 1993. Soil micromorphology. In Whittle, A., Rouse, A.J. and Evans, J.G. (eds.) A Neolithic downland monument in its environment: excavations at the Easton Down long barrow, Bishops Canning, north Wiltshire, 218-219, 234-235. *Proceedings of the Prehistoric Society* 59, 197-239.

Macphail, R.I. 1994a. The reworking of urban stratigraphy by human and natural processes. In Hall, A.R. and Kenward, H.K. (eds.) *Urban-rural*

Connexions: Perspectives from Environmental Archaeology, 13-43. Oxford: Oxbow.

Macphail, R.I. 1994b. Soil micromorphology investigations in archaeology, with special reference to drowned coastal sites in Essex. *Seesoil* 10, 13-28

Macphail, R.I. 1998a. A reply to Carter and Davidson's 'An evaluation of the contribution of soil micromorphology to the study of ancient arable agriculture'. *Geoarchaeology* 13(6), 549-64.

Macphail, R.I. 1998b. Discussion and update. *Bulletin of the Archaeological Soil Micromorphology Working Group* 1. Report on workshop at the Institute of Archaeology/Department of Geology, University College London, March 1997.

Macphail, R.I. and Crowther, J. 2008. Soil micromorphology, chemistry and magnetic susceptibility. In Cooke, N., Brown, F. and Phillpotts, C. (eds.) *From Hunter Gatherers to Huntsmen*, CD-Rom Chapter 30. Salisbury: Framework Archaeology.

Macphail, R.I. and Crowther, J., 2009. Soil micromorphology, chemistry and magnetic susceptibility. In Phillips, M., Duncan, H. and Mallows, C. (eds.) *Four Millennia of Human Activity Along the A505 Baldock Bypass, Hertfordshire*, 84-85, 122-123, CD-ROM Appendix 4. Bedford: Albion Archaeology.

Macphail, R.I. and Cruise, G.M. 1996. Soil micromorphology. In Bell, M., Fowler, P.J. and Hillson, S.W. (eds.) *The Experimental Earthwork Project 1960-1992*, 95-107. Council for British Archaeology Research Report 100. York: Council for British Archaeology.

Macphail, R.I and Goldberg, P. 1990. The micromorphology of tree subsoil hollows: their significance to soil science and archaeology. In Douglas, L.A. (ed.) *Soil Micromorphology: A Basic and Applied Science. Proceedings of the 8th International Working Meeting of Soil Micromorphology, San Antonio, Texas, July 1988*, 425-429. Developments in Soil Science 19. Amsterdam: Elsevier Science Publishers B.V.

Macphail, R.I. and Goldberg, P. 1995. Recent advances in micromorphological interpretations of soils and sediments from archaeological sites. In Barham, A.J. and Macphail, R.I. (eds.) *Archaeological Sediments and Soils: Analysis, Interpretation and Management*, 1-24. London: Institute of Archaeology, University College London.

Macphail, R.I., Cammas, C., Gebhardt, A., Langohr, R. and Linderholm, J. 1998. Anthropogenic influences on soils in the Late Quaternary. *Proceedings of the 16th World Congress of Soil Science, Montpellier, August 1998. Symposium 16. Records in Soils of Environmental and Anthropogenic Changes*, on CD-ROM. Montpellier: International Society of Soil Science.

Macphail, R.I., Courty, M.A. and Gebhardt, A. 1990. Soil micromorphological evidence of early agriculture in north-west Europe. *World Archaeology* 22(1), 53-69. (Cited as Macphail *et al.* 1990a)

Macphail, R.I., Courty, M.A. and Goldberg, P. 1990. Soil micromorphology in archaeology. *Endeavour, New Series* 14(4), 163-71. (Cited as Macphail *et al.* 1990b)

Macphail, R.I., Cruise, G.M., Engelmark, R. and Linderholm, J. 2000. Integrating soil micromorphology and rapid chemical survey methods: new developments in reconstructing past rural settlement and landscape organization. In Roskams, S. (ed.) *Interpreting Stratigraphy*, Volume 9, 71-80. York: University of York.

Macphail, R.I., Cruise, G.M., Allen, M.J., Linderholm, J. and Reynolds, P. 2004. Archaeological soil and pollen analysis of experimental floor deposits; with special reference to Butser Ancient Farm, Hampshire, UK. *Journal of Archaeological Science* 31:175-91.

Macphail, R.I., Galinie, H. and Verhæge, F. 2003. A future for dark earth? *Antiquity* 77(296), 349-58.

Macphail, R.I., Romans, J.C.C. and Robertson, L. 1987. The application of micromorphology to the understanding of Holocene soil development in the British Isles; with special reference to early cultivation. In Federoff, N., Bresson, L.M. and Courty, M.A. (eds.) *Soil Micromorphology. Proceedings of the 7th International Working Meeting on Soil Micromorphology, Paris – July 1985*, 647-56. Plaisir: L'Association Français pour l'Étude du Sol.

Madella, M. 1997. Phytoliths from a central Asia loess-palaeosol sequence and modern soils: their taphonomical and palaeoecological implications. In Pinilla, A., Juan-Tresserras, J. and Machado, M.J. (eds.) *Estado Actual de los Estudios de Fitolitos en Suelos y Plantas*, 49-57. Consejo Superior de Investigaciones Científicas Monografias 4. Madrid: Centro de Ciencias Medoambientales del Consjo Superior de Investigaciones Científicas.

Madsen, P.K. 1980. Medieval ploughing marks in Ribe. *Tools and Tillage* 4(1), 36-45.

Maltby, E. and Caseldine, C.J. 1982. Prehistoric soil and vegetation development on Bodmin Moor, southwestern England. *Nature* 297, 397-400.

Mann, L.K. 1985. A regional comparison of carbon in cultivated and uncultivated alfisols and mollisols in the central United States. *Geoderma* 36(3-4), 241-53.

Martin, T.C., Robertson, J.A. and Pawluk, S. 1987. Changes in the micromorphology and analytical properties of a grey luvisol (Cryoboralf) under a 49 year old forage stand. *Geoderma* 40(1-2), 209-24.

McGlade, J. 1995. Archaeology and the ecodynamics of human-modified landscapes. *Antiquity* 69, 113-32.

McIntyre, D.S. 1958. Soil splash and the formation of surface crusts by raindrop impact. *Soil Science* 85, 261-66.

McKeague, J.A. 1983. Clay skins and argillic horizons. In Bullock, P. and Murphy, C.P. (eds.) *Soil Micromorphology Volume 2: Soil Genesis. Proceedings of the 6th International Working Meeting on Soil Micromorphology, London 198* 367-388. Berkhamsted: AB Academic Publishers.

McKeague, J.A., Fox, C.A., Stone, J.A. and Protz, R. 1987. Effects of cropping system on structure of Brookston clay loam in long-term experimental plots at Woodslee, Ontario. *Canadian Journal of Soil Science* 67, 571-84.

Mercer, R. 1970. The Neolithic settlement on Carn Brea: preliminary report, 1970. *Cornish Archaeology* 9, 53-62.

Mercer, R. 1981a. Introduction. In Mercer, R. (ed.) *Farming Practice in British Prehistory,* ix-xxvi. Edinburgh: Edinburgh University Press.

Mercer, R. 1981b. Excavations at Carn Brea, Illogan, Cornwall, 1970-73. A Neolithic fortified complex of the 3rd millennium bc. *Cornish Archaeology* 20, 1-204.

Mermut, A.R. and St. Arnaud, R.J. 1981. Microband fabric in seasonally frozen soils. *Journal* of *the Soil Science Society of America* 45, 578-86.

Meurers-Balke, J. 1985. Experimente zum Anbau und zur Verarbeiterung prähistorischer Getreidearten. *Archäologische Informationen* tome 8, cahier 1, 8-17.

Middleton, W.D. and Price, T.D. 1996. Identification of activity areas by multi-element characterization of sediments from modern and archaeological house floors using Inductively Coupled Plasma-atomic Emission Spectroscopy. *Journal of Archaeological Science* 23, 673-87.

Miedema, R. and van Oort, F. 1990. Significance of soil microfabric for soil physical characteristics and behaviour of late Weichelian and Holocene Rhine deposits in the Netherlands. In Douglas, L.A. (ed.) *Soil Micromorphology: A Basic and Applied Science. Proceedings of the 8th International Working Meeting of Soil Micromorphology, San Antonio, Texas - July 1988*, 23-31. Developments in Soil Science 19. Amsterdam: Elsevier Science Publishers B.V.

Mikkelsen, J.H. and Langohr, R. 1996. A pedological characterization of the Aubechies soil, a well preserved soil sequence dated to the earliest Neolithic agriculture in Belgium. In Castelletti, L. and Cremashi, M. (eds.) *Paleoecology. XIIIth International Congress of Prehistoric and Protohistoric Sciences, Forlì (Italia), 8/14 September 1996. Colloquia: V. Numerical analysis and image processing in archaeobotany; VI. Micromorphology of deposits of anthropogenic origin*, 143-49. Forlì: A.B.A.C.O.

Mitchell, G.F. 1972. Soil deterioration associated with prehistoric agriculture in Ireland. *Proceedings of the 24th International Geological Congress Symposium, Montréal* 1, 59-68.

Moffat, A.J. 1988. The distribution of 'Celtic fields' on the East Hampshire chalklands. *Proceedings of the Hampshire Field Club Archaeological Society* 44, 11-23.

Moore, P.D. 1982. Sub-surface formation of charcoal: an unlikely event in peat. *Quaternary Newsletter* 38, 13-14.

Morrison, K.D. 1994. The intensification of production: archaeological approaches. *Journal of Archaeological Method and Theory* 1(2), 111-59.

Mrozowski, S.A. 1994. The discovery of a Native American cornfield on Cape Cod. *Archaeology of Eastern North America* 22, 47-62.

Mücher, H.J. and de Ploey. 1977. Experimental and micromorphological investigation of erosion and redeposition of loess by water. *Earth Surface Processes and Landforms* 2, 117-24.

Mücher, H.J. and Morozova, T.D. 1983. The application of soil micromorphology in Quaternary geology and geomorphology. In Bullock, P. and Murphy, C.P. (eds.) *Soil Micromorphology, Volume 1: Techniques and Applications. Proceedings of the 6th International Working Meeting on Soil Micromorphology, London 1981,* 151-94. Berkhamsted: AB Academic Publishers.

Murphy, P.M. 1992. Chapter 8: Environmental studies: Culver Street. In Crummy, P. *Excavations at Culver Street, the Gilberd School, and Other Sites in Colchester 1971-85,* 273-87. Colchester Archaeological Report 6. Colchester: Colchester Archaeological Trust Ltd. and English Heritage.

Nascimento, V.M., Almendros, G. and Fernandes, F.M. 1992. Soil humus characteristics in virgin and cleared areas of the Paraná River basin in Brazil. *Geoderma* 54, 137-50.

Netting, R. McC. 1993. *Smallholders, Householders: Farm Families and the Ecology of Intensive,*

Sustainable Agriculture. Stanford: Stanford University Press.

Nettleton, W.D., Eswaran, H., Holzhey, C.S. and Nelson, R.E. 1987. Micromorphological evidence of clay translocation in poorly dispersible soils. *Geoderma* 40(1-2), 37-48.

Nielsen, V. 1970. Iron Age plough-marks in Store Vildmose, north Jutland. *Tools and Tillage* 1(3), 151-65.

Nielsen, V. 1984. Prehistoric field boundaries in eastern Denmark. *Journal of Danish Archaeology* 3, 135-63.

Nielsen, V. 1986. Ploughing in the Iron Age. Plough marks in Store Vildmose, North Jutland. *Journal of Danish Archaeology* 5, 189-208.

Nielsen, V. 1993. *Jernalderens Pløjning. Store Vildmose.* Hjørring: Vendsyssel Historiske Museum.

Noe, P. 1976. Pre-Medieval plough marks in Viborg. *Tools and Tillage* 3(1), 59-64.

Norton, L.D. 1987. Micromorphological study of surface seals developed under simulated rainfall. *Geoderma* 40(1-2), 127-40.

Oades, J.M. 1993. The role of biology in the formation, stabilisation and degradation of soil structure. *Geoderma* 56, 377-400.

Odell, G.H. and Cowan, F. 1987. Estimating tillage effects on artifact distributions. *American Antiquity* 52, 456-84.

Olson, G.W. 1981. Archaeology: lessons on future soil use. *Journal of Soil and Water Conservation* 36(5), 261-64.

Pagliai, M. 1987. Effects of different management practices on soil structure and surface crusting. In Federoff, N., Bresson, L.M. and Courty, M.-A. (eds.) *Soil Micromorphology. Proceedings of the 7th International Working Meeting on Soil Micromorphology, Paris - July 1985*, 415-22. Plaisir: Association Française pour l'Etude du Sol.

Pagliai, M. 1994. Micromorphology and soil management. In Ringrose-Voase, A.J. and Humphreys, G.S. (eds.) *Soil Micromorphology: Studies in Management and Genesis. Proceedings of the 9th International Working Meeting on Soil Micromorphology, Townsville, Australia, July 1992*, 623-40. Developments in Soil Science 22. Amsterdam: Elsevier.

Pagliai, M. and De Nobili, M. 1993. Relationships between soil porosity, root development and soil enzyme activity in cultivated soils. *Geoderma* 56, 243-56.

Pagliai, M. and La Marca, M. 1979. Micromorphological study of soil crusts. *Agrochimica* 23, 16-25.

Pagliai, M., Bisdom, E.B.A. and Ledin, S. 1983. Changes in surface structure (crusting) after application of sewage sludges and pig slurry to cultivated agricultural soils in northern Italy. *Geoderma* 30, 35-53.

Pagliai, M., La Marca, M., Lucamante, G. and Genovese, L. 1984. Effects of zero and conventional tillage on the length and irregularity of elongated pores in a clay loam soil under viticulture. *Soil and Tillage Research* 4, 433-44.

Pagliai, M., Pezzarossa, B., Mazzoncini, M. and Bonari, E. 1989. Effects of tillage on porosity and microstructure of a loam soil. *Soil Technology* 2, 345-58.

Pals, J.P. and van Dierendonck, M.C. 1988. Between flax and fabric: cultivation and processing of flax in a Mediaeval peat reclamation settlement near Midwoud (Prov.-Noord Holland). *Journal of Archaeological Science* 15, 237-51.

Pape, J.C. 1970. Plaggen soils in the Netherlands. *Geoderma* 4, 229-56.

Parry, M.L. 1976. A typology of cultivation ridges in southern Scotland. *Tools and Tillage* 3(1), 3-19.

Patterson, W.A. III, Edwards, K.J. and Maguire, D.J. 1987. Microscopic charcoal as a fossil indicator of fire. *Quaternary Science Reviews* 6, 3-23.

Pawluk, S. 1980. Micromorphological investigations of cultivated Gray Luvisols under different management practices. *Canadian Journal of Soil Science* 60, 731-45.

Payne, F.G. 1947. The plough in ancient Britain. *The Archaeological Journal* 104(3), 82-111.

Payne, P.C.J. and Tanner, D.W. 1959. The relationship between rake angle and the performance of simple cultivation implements. *Journal of Agricultural Research* 4(4), 312-25.

Pearsall, D.M. 1978. Phytolith analysis of archaeological soils: evidence for maize cultivation in Ecuador. *Science* 199, 177-78.

Pearsall, D.M. and Trimble, M.K. 1984. Identifying past agricultural activity through soil phytolith analysis: a case study from the Hawaiian Islands. *Journal of Archaeological Science* 2, 119-33.

Peterson, J.W.M. 1990. Why did the idea of coaxial field systems last so long? *Antiquity* 64(244), 584-91.

Pitty, A.F. 1978. *Geography and Soil Properties*. London: Methuen.

Prøsch-Danielsen, L. and Simonsen, A. 1988. Principal components analysis of pollen, charcoal and soil phosphate data as a tool in prehistoric landuse investigations at Forsandmoen, southwest Norway. *Norwegian Archaeological Review* 21(2), 85-102.

Proudfoot, V.B. 1958. Problems of soil history. Podzol development at Goodland and Torr Townlands, Co. Antrim, Northern Ireland. *Journal of Soil Science* 9(2), 186-98.

Provan, D.M.J. 1971. Soil phosphate analysis as a tool in archaeology. *Norwegian Archaeological Review* 4, 37-50.

Provan, D.M.J. 1973. The soils of an Iron Age farm site – Bjellandsøynæ, SW Norway. *Norwegian Archaeological Review* 6, 30-41.

Pryor, F. 1976. Fen-edge land management in the Bronze Age: an interim report on excavations at Fengate, Peterborough, 1971-75. In Burgess, C. and Miket, R. (eds.) *Settlement and Economy in the Third and Second Millennia BC*, 29-49. Oxford: British Archaeological Reports.

Pryor, F. 1995. Abandonment and the role of ritual sites in the landscape. *Scottish Archaeological Review* 9-10, 96-109.

Quine, T.A. 1995. Soil analysis and archaeological site formation studies. In Barham, A.J. and Macphail, R.I. (eds.) *Archaeological Sediments and Soils: Analysis, Interpretation and Management*, 77-98. London: Institute of Archaeology, University College London.

Ramskou, Th. 1981. Lindholm Høje. A Danish Viking period field. *Tools and Tillage* 4(2), 98-109.

Rees, S. 1979a. Stone ard points from Orkney and Shetland. *Tools and Tillage* 3(4), 249-54.

Rees, S. 1979b. *Agricultural Implements in Prehistoric and Roman Britain, Parts i and ii*. British Archaeological Reports British Series 69 (i and ii). Oxford: British Archaeological Reports.

Rees, S. 1981. Agricultural tools: function and use. In Mercer, R. (ed.) *Farming Practice in British Prehistory*, 66-84. Edinburgh: Edinburgh University Press.

Reid, I., Parkinson, R., Twomlow, S. and Clark, A. 1990. The impact of agricultural landuse changes on soil conditions and drainage. In Thornes, J.B. (ed.) *Vegetation and Erosion: Processes and Environments*, 199-215. Chichester: John Wiley and Sons.

Reynolds, K.S. and Catt, J.A. 1987. Soils and vegetation history of abandoned enclosures in the New Forest, Hampshire, England. *Journal of Archaeological Science* 147, 507-27.

Reynolds, P.J. 1967. Experiments in Iron Age Agriculture. *Transactions of the Bristol and Gloucestershire Archaeological Society* 86, 60-73.

Reynolds, P.J. 1977. Slash and burn experiment. *Archaeological Journal* 134, 307-18.

Reynolds, P.J. 1979. *Iron-age Farm: The Butser Experiment*. London: British Museum Publications Limited.

Reynolds, P.J. 1981. Deadstock and livestock. In Mercer, R. (ed.) *Farming Practice in British Prehistory*, 97-122. Edinburgh: Edinburgh University Press.

Reynolds, P.J. 1982. The ploughzone. *Festschrift zum 100 jahrigen Jubitaun der Abteilung der Naturhistorischen Gesellschaft*, 315-41. Nurnberg E.V.

Reynolds, P.J. 1987. *Butser Ancient Farm Yearbook 1986*. Horndean: Butser Ancient Farm Project Trust.

Reynolds, P.J. 1988. *Butser Ancient Farm Yearbook 1987*. Horndean: Butser Ancient Farm Project Trust.

Reynolds, P.J. 1989. *Butser Ancient Farm Year Book 1988*. Horndean: Butser Ancient Farm Project.

Ringrose-Voase, A.J. 1991. Micromorphology of soil structure: description, quantification, application. *Australian Journal of Soil Research* 29, 777-813.

Roberts, B.K. 1987. Landscape archaeology. In Wagstaff, J.M. (ed.) *Landscape and Culture*, 77-95. Oxford: Basil Blackwell.

Roberts, B.K. 1996. *Landscapes of Settlement: Prehistory to the Present*. London and New York: Routledge.

Robinson, M. and Wilson, B. 1987. A survey of the environmental archaeology in the south midlands. In Keeley, H.C.M. (ed.) *Environmental Archaeology: A Regional Review. Volume II*, 16-100. Historic Buildings and Monuments Commission for England Occasional Paper No. 1. London: Historic Buildings and Monuments Commission for England.

Romans, J.C.C. 1986. Some notes on the soils at Scord of Brouster, Shetland. In Whittle, A., Keith-Lucas, M., Milles, A., Noddle, B., Rees, S. and Romans, J.C.C. *Scord of Brouster. An early agricultural settlement on Shetland. Excavations 1977-1979*, 125-31. Oxford

University Committee for Archaeology Monograph No. 9. Oxford: Oxford Committee for Archaeology.

Romans, J.C.C. and Robertson, L. 1975. Soils and archaeology in Scotland. In Evans, J. G., Limbrey, S. and Cleere H. (eds.) *The Effect of Man on the Landscape: the Highland Zone*, 37-9. Council for British Archaeology Research Report No. 11. London: Council for British Archaeology.

Romans, J.C.C. and Robertson, L. 1983a. The general effects of early agriculture on the soil profile. In Maxwell, G.S. (ed.) *The Impact of Aerial Reconnaissance on Archaeology*, 136-41. Council for British Archaeology Research Report No. 49. London: Council for British Archaeology.

Romans, J.C.C. and Robertson, L. 1983b. The environment of north Britain: soils. In Chapman, J.C. and Mytum, H.C. (eds.) *Settlement in North Britain 1000 BC to AD 1000. Papers presented to George Jobey, Newcastle-upon-Tyne, December 1982*, 55-80. British Archaeological Reports British Series 118. Oxford: British Archaeological Reports.

Romans, J.C.C., Durno, S.E. and Robertson, L. 1973. A fossil brown forest soil from Angus. *Journal of Soil Science* 24, 125-28.

Roper, D.C. 1976. Lateral displacement of artifacts due to plowing. *American Antiquity* 41, 372-75.

Rowley-Conwy, P. 1981. Slash and burn in the temperate European Neolithic. In Mercer, R. (ed.) *Farming Practice in British Prehistory*, 85-96. Edinburgh: Edinburgh University Press.

Rowley-Conwy, P. 1987. The interpretation of ard marks. *Antiquity* 61, 263-66.

Runnels, C.N. 1995. Environmental degradation in ancient Greece. *Scientific American* 272, 72-75.

Russell, K.W. 1988. *After Eden: The Behavioural Ecology of Early Food Production in the Near East and North Africa.* British Archaeological Reports International Series 391. Oxford: British Archaeological Reports.

Ruthenberg, H. 1980. *Farming Systems in the Tropics*. 3rd edition. Oxford: Clarendon Press.

Šach, F. 1968. Proposal for the classification of pre-industrial implements. *Tools and Tillage* 1(1), 3-27.

Sandor, J.A. 1992. Long-term effects of prehistoric agriculture on soils: examples from New Mexico and Peru. In Holliday, V.T. (ed.) *Soils in Archaeology: Landscape Evolution and Human Occupation,* 217-45. Washington and London: Smithsonian Institution Press.

Sandor, J.A. and Eash, N.S. 1991. Significance of ancient agricultural soils for long-term agronomic studies and sustainable agriculture research. *Agronomy Journal* 83, 29-37.

Santiago Mora, C., Herrera, L.F., Inés Cavelier, F. and Rodríguez, C. 1991. *Cultivars, Anthropic Soils and Stability: A Preliminary Report of Archaeological Research in Araracuara, Colombian Amazonia.* (Translated by N. Verniquet) University of Pittsburgh Latin American Archaeology Reports No. 2. Pittsburgh: Department of Anthropology, University of Pittsburgh.

Schafer, R.L. and Johnson, C.E. 1982. Changing soil condition - the soil dynamics of tillage. In *Predicting Tillage Effects on Soil Physical Properties and Processes*, 13-28. American Society of Agronomy Special Publication Number 44. Madison: American Society of Agronomy and Soil Science Society of America.

Schama, S. 1995. *Landscape and Memory*. London: Harper Collins.

Schimel, D.S., Coleman, D.C. and Horton, K.A. 1985. Soil organic matter dynamics in paired rangeland and cropland toposequences in North Dakota. *Geoderma* 36(3-4), 201-14.

Schofield, A.J. 1990. *Interpreting Artefact Scatters: Contributions to Ploughzone Archaeology*. Oxford: Oxbow Books.

Scharpenseel, H.W. and Becker-Heidmann, P. 1992. Twenty-five years of radiocarbon dating soils: paradigm of erring and learning. *Radiocarbon* 34, 541-49.

Sherratt, A. 1981. Plough and pastoralism: aspects of the secondary products revolution. In Hodder, I., Isaac, G. and Hammond, N. (eds.) *Pattern of the Past: Studies in Honour of David Clarke*, 261-306. Cambridge: Cambridge University Press.

Sherratt, A. 1987. Wool, wheels and ploughmarks: local developments or outside introductions in Neolithic Europe? *Bulletin of the Institute of Archaeology* 23, 1-15.

Shiel, R. and Askew, G.P. 1988. Description of thin sections taken from plaggen layers at Iona. *Proceedings of the Society of Antiquaries of Scotland* 118, fiche 3, A5-14.

Shipitalo, M.J. and Protz, R. 1989. Chemistry and micromorphology of aggregation in earthworm casts. *Geoderma*, 45(3-4), 357-74.

Sibbesen, E., Andersen, C.E., Andersen, S. and Flensted-Jensen, M. 1985. Soil movement in long-term field experiments as a result of cultivations. I. A model for approximating soil movement in one horizontal dimension by repeated tillage. *Experimental Agriculture* 21, 101-7.

Simpson, I.A. 1985. Stable carbon isotope analysis of anthropogenic soils and sediments in Orkney. In Fieller, N.R.J., Gilbertson, D.D. and Ralph, N.G.A. (eds.) *Palaeoenvironmental Investigations – Research Design, methods and Data Analysis*, 55-66. British Archaeological Reports, International Series 258. Oxford: British Archaeological Reports.

Simpson, I.A. 1993. The chronology of anthropogenic soil formation in Orkney. *Scottish Geographical Magazine* 109, 4-11.

Simpson, I.A. 1994. Spatial constraints on anthropogenic soil formation in Orkney. *Scottish Geographical Magazine* 110, 184-86.

Simpson, I.A. 1995. Establishing time-scales for early cultivated soils in the Northern Isles of Scotland. *Scottish Geographical Magazine* 111, 184-86.

Simpson, I.A. 1997. Relict properties of anthropogenic deep top soils as indicators of infield management in Marwick, West Mainland, Orkney. *Journal of Archaeological Science* 24, 365-80.

Simpson, I.A., Dockrill, S.J., Bull, I.D. and Evershed, R.P. 1998. Early anthropogenic soil formation at Tofts Ness, Sanday, Orkney. *Journal of Archaeological Science* 25, 729-46.

Smart, T.L. and Hoffman, E.S. 1988. Environmental interpretation of archaeological charcoal. In Hastorf, C. and Popper, V.S. (eds.) *Current Paleoethnobotany*, 167-205. Chicago and London: University of Chicago Press.

Smithson, F. 1958. Grass opal in British soils. *Journal of Soil Science* 9(1), 148-54.

Soto, B., Basanta, R., Perez, R., Díaz-Fierros, F. 1995. An experimental study of the influence of traditional slash-and-burn practices on soil erosion. *Catena* 24, 13-23.

Spek, T. 1992. The age of plaggen soils. In Verhoeve, A. and Vervloet, J.A.J. (eds.) *The Transformation of the European Rural Landscape: Methodological Issues and Agrarian Change*, 72-91. Brussels: NFWO–FNRS.

Spoor, G. 1969a. Design of soil engaging implements. *Farm Machine Design Engineering*, September 1969, 22-25.

Spoor, G. 1969b. Design of soil engaging implements 2. Farm Machine Design Engineering, December 1969, 14-9.

Spoor, G. 1975. Fundamental aspects of cultivation and implement selection. In *Soil Physical Conditions and Crop Production*, 128-44. Ministry of Agriculture, Fisheries and Food Technology Bulletin 29. London.

Spoor, G. and Godwin, R.J. 1978. An experimental investigation into the deep loosening of soil by rigid tines. *Journal of Agricultural Engineering Research* 23, 243-58.

Spratt, D.A. 1978. Prehistoric field and land boundary systems in the North York Moors. In Bowen, H.C. and Fowler, P.J. (eds.) *Early Land Allotment*, 115-18. British Archaeological Reports, British Series 48. Oxford: British Archaeological Reports.

Steensberg, A. 1936. North west European plough types of prehistoric times and the middle ages. *Acta Archaeologica* 7, 244-80.

Steensberg, A. 1955. Mit braggender Flame. *KUML* 1955, 63-103.

Steensberg, A. 1957. Some recent experiments in neolithic agriculture. *Agricultural History Review*, 5, 66-73.

Steensberg, A. 1957-8. Parallel ploughing with alternative sloping and upright ard in Columella. *Folk-Living*, 1957-8, 157-162.

Steensberg, A. 1971-2. The practice of tilling spades in Asia and Europe. Off-print from *A Magyar Mezõgazdasági Múzeum Közleményei, 1971-1972.* Budapest.

Steensberg, A. 1973. A 6,000 year old ploughing implement from Satrup Moor. *Tools and Tillage* 2(2), 105-18.

Steensberg, A. 1977. Stone shares of ploughing implements from the Bronze Age of Syria. *Det Kongelige Danske Videnskabernes Selskab Historisk-filosofiske Meddelelser* 47/6.

Steensberg, A. 1979. *Draved: An Experiment in Stone Age Agriculture: Burning, Sowing and Harvesting.* Copenhagen: The National Museum of Denmark.

Steensberg, A. 1986. *Man the Manipulator. An Ethno-archaeological Basis for Reconstructing the Past.* Copenhagen: The National Museum of Denmark.

Stevens, C.J. 1996. *Iron Age and Roman Agriculture in the Upper Thames Valley: Archaeobotanical and Social Perspectives.* Unpublished Ph. D. thesis, University of Cambridge.

Studman, C.J. and Field, J.E. 1975. The motion of a stone embedded in non-cohesive soil disturbed by a moving tine. *Journal of Terramechanics* 12(3-4), 131-47.

Sutter, J.W. 1987. Cattle and inequality: herd size differences and pastoral production among the Fulani of northeastern Senegal. *Africa* 57(2), 196-218.

Swain, A.M. 1973. A history of fire and vegetation in northeastern Minnesota as recorded in lake sediments. *Quaternary Research* 3, 76-86.

Tanner, D.W. 1960. Further work on the relationship between rake angle and the performance of simple cultivation implements. *Journal of Agricultural Engineering Research* 5(3), 307-15.

Tarchitzky, J., Banin, A., Morin, J. and Chen, Y. 1984. Nature, formation and effects of soil crust formed by water drop impact. *Geoderma* 33, 135-55.

Tarlow, S. 1990. *Metaphors and Neolithic/Bronze Age Burial Mounds*. Unpublished M.Phil. dissertation, University of Cambridge.

Tarlow, S. 1993/4. Scraping the bottom of the barrow: an agricultural metaphor in Neolithic/Bronze-Age European burial practice. *Journal of Theoretical Archaeology* 3(4), 123-44.

Taylor, C. 1975. *Fields in the English Landscape*. London: J.M. Dent and Sons Ltd.

Taylor, C. 1981. Archaeology and the origins of open-field agriculture. In Rowley, T. (ed.) *The Origins of Open-field Agriculture*, 13-21. London: Croom Helm.

Terwilliger, V.J. 1990. Effects of vegetation on soil slippage by pore pressure modification. *Earth Surface Processes and Landforms* 15(6), 553-70.

Thomas, C. 1970. Bronze Age spade marks at Gwithian, Cornwall. In Gailey, A. and Fenton, A. (eds.) *The Spade in Northern and Atlantic Europe*, 10-17. Holywood and Belfast: Ulster Folk Museum and the Institute of Irish Studies, Queen's University.

Thomas, C. 1978. Types and distributions of pre-Norman fields in Cornwall and Scilly. In Bowen, H.C. and Fowler, P.J. (eds.) *Early Land Allotment*, 7-15. British Archaeological Reports, British Series 48. Oxford: British Archaeological Reports.

Thomas, J. 1991. *Rethinking the Neolithic*. Cambridge: Cambridge University Press.

Thomas, J. 1994. Pict's Knowe, Dumfries, Excavations 1994: Interim Report. http://avebury.arch.soton.ac.uk/Research/PictsKnowe_1994.

Thomas, J. 1996. Neolithic houses in mainland Britain: a sceptical view. In Darvill, T. and Thomas, J. (eds.) *Neolithic Houses in Northwest Europe and Beyond*, 1-12. Oxbow Monograph 57. Oxford: Oxbow Monographs).

Thomas, J. 1999. *Understanding the Neolithic*. London and New York: Routledge.

Thomas, K. 1990. Aspects of soils and early agriculture. *World Archaeology* 22, vii-xiii.

Thompson, M.L., Federoff, N. and Fournier, B. 1990. Morphological features related to agriculture and faunal activity in three loess-derived soils in France. *Geoderma* 46(4), 329-49.

Thompson, M.L., Singh, P., Corak, S. and Straszheim, W.E. 1992. Cautionary notes for the automated analysis of soil pore-space images. *Geoderma* 55, 159-74.

Thompson, T.R.E. 1983. Translocation of fine earth in some soils from an area in mid Wales. In Bullock, P. and Murphy, C.P. (eds.) *Soil Micromorphology II. Soil Genesis. Proceedings of the 6th International Working Meeting on Soil Micromorphology, London 1981,* 531-39. Berkhamsted: AB Academic Publishers.

Thornes, J.B. 1987. The palaeo-ecology of erosion. In Wagstaff, J.M. (ed.) *Landscape and Culture* 37-56. Oxford: Basil Blackwell.

Tipping, R. 1996. Microscopic charcoal records, inferred human activity and climate change in the Mesolithic of northernmost Scotland. In Pollard, T. and Morrison, A. (eds.) *The Early Prehistory of Scotland*, 39-61. Edinburgh: Edinburgh University Press.

Tipping, R., Carter, S. and Johnston, D. 1994. Soil pollen and soil micromorphological analyses of old ground surfaces on Biggar Common, Borders Region, Scotland. *Journal of Archaeological Science* 21, 387-401.

Turelle, J.W. and McCalla, T.M. 1961. Photomicrographic study of soil aggregates and microorganisms as influenced by stubble mulching and ploughing. *Soil Science Society of America Proceedings* 25, 487-90.

Tvengsberg, P.M. 1996. Rye and swidden cultivation tillage without tools. *Tools and Tillage* 7(4), 131-46.

Ullrich, J.M. 2010. *Geochemical contributions to the arcaeological examination of promontory forts: the use of phosphate analysis techniques on Achillbeg and Achill Islands, Co. Mayo.* Unpublished PhD thesis. University College Dublin.

Usai, M.-R. 2001. Textural pedofeatures and pre-Hadrian's Wall ploughed paleosols at Stanwix, Carlisle, Cumbria, UK. *Journal of Archaeological Science* 28, 541-53

Valentin, C. and Bresson, L.-M. 1992. Morphology, genesis and classification of surface crusts in loamy and sandy soils. *Geoderma* 55, 225-45.

Valentin, C. and Ruiz Figueroa, J.F. 1987. Effects of kinetic energy and water application rate on the development of crusts in a fine sandy loam soil using sprinkling irrigation and rainfall simulation. In Fedoroff, N., Bresson, L.M. and Courty, M.A. (eds.) *Soil Micromorphology. Proceedings of the 7th International Working Meeting on Soil Micromorphology, Paris - July 1985*, 401-8. Plaisir: l'Association Française pour l'Étude du Sol.

van de Westeringh, W. 1988. Man-made soils in the Netherlands, especially in sandy areas ('Plaggen soils'). In Groenman-van Waateringe, W. and Robinson, M. (eds.) *Man-made Soils*, 5-20. British Archaeological Reports, International Series 410. Oxford: British Archaeological Reports.

van Vliet-Lanoë, B. 1985a. Frost effects in soils. In Boardman, J. (ed.) *Soils and Quaternary Landscape Evolution*, 117-58. Chichester: J. Wiley and Sons Ltd.

van Vliet-Lanoë, B. 1985b. From frost to gelifluction: a new approach based on micromorphology. Its applications to arctic environment. *Inter-Nord,* 17, 15-20.

van Vliet-Lanoë, B., Helluin, M. Pellerin, J. and Valadas, B. 1992. Soil erosion in western Europe: from the last interglacial to the present. In Bell, M. and Boardman, J. (eds.) *Past and Present Soil Erosion: Archaeological and Geological Perspectives*, 101-14. Oxbow Monograph 22. Oxford: Oxbow Books.

Verhaegen, Th. 1984. The influence of soil properties on the erodibility of Belgian loamy soils: a study based on rainfall simulation experiments. *Earth Surface Processes and Landforms* 9, 499-507.

Verrill, L. and Tipping, R. 2010. Use and abandonment of a Neolithic field system at Belderrig, Co. Mayo, Ireland: Evidence for economic marginality. *The Holocene* 20(7), 1011-21.

Vogel, H.J., Weller, U. and Babel, U. 1993. Estimating orientation and width of channels and cracks at soil polished blocks - a stereological approach. *Geoderma* 56, 301-16.

Vuorela, I. 1976. An instance of slash and burn cultivation in northern Finland investigated by pollen analysis of a mineral soil. *Memoranda Societas Fauna Flora Fennica* 52, 29-46.

Wagstaff, J.M. 1987. Introduction. In Wagstaff, J.M. (ed.) *Landscape and Culture*, 1-10. Oxford: Basil Blackwell.

Washburn, A.L. 1979. *Geocryology: A Survey of Periglacial Processes and Environments*. Second edition. London: Edward Arnold.

Wattez, J. and Courty, M.A. 1987. Morphology of ash of some plant materials. In Federoff, N., Bresson, L.M. and Courty, M.A. (eds.) *Soil Micromorphology. Proceedings of the 7th International Working Meeting on Soil Micromorphology, Paris - July 1985*, 677-683. Plaisir: L'Association Français pour l'Étude du Sol.

Whittle, A. 1977. *The Earlier Neolithic of Southern England and its Continental Background.* British Archaeological Reports Supplementary Series 35. Oxford: British Archaeological Reports.

Whittle, A., Keith-Lucas, M., Milles, A., Noddle, B., Rees, S. and Romans, J.C.C. 1986. *Scord of Brouster. An early agricultural settlement on Shetland. Excavations 1977-1979.* Oxford University Committee for Archaeology Monograph No. 9. Oxford: Oxford Committee for Archaeology.

Whittle, A., Rouse, A.J. and Evans, J.G. (eds.) 1993. A Neolithic downland monument in its environment: excavations at the Easton Down long barrow, Bishops Canning, north Wiltshire. *Proceedings of the Prehistoric Society* 59, 197-239

Wickstead, H. 2007. *Land Division and Identity in Later Prehistoric Dartmoor, South-West Britain: Translocating Tenure.* Unpublished PhD thesis. University College London.

Wieder, M. and Yaalon, D.H. 1978. Grain cutans resulting from clay illuviation in calcareous soil material. In Delgado, M. (ed.) *Micromorfologia de Suelos*, 1133-1158. Granada: Universidad de Granada.

Wild Heerbrugg Ltd. 1978. *Photomakroskop Wild M400: Instructions for Use*. Heerbrugg: Wild Heerbrugg.

Wilkinson, T.J. 1982. The definition of ancient manured zones by means of extensive sherd-sampling techniques. *Journal of Field Archaeology* 9, 321-33.

Wilkinson, T.J. 1989. Extensive sherd scatters and land-use intensity: some recent results. *Journal of Field Archaeology* 16, 31-46.

Williams, P.J. and Smith, M.W. 1989. *The Frozen Earth: Fundamentals of Geocryology*. Cambridge: Cambridge University Press.

Wintle, A.G. and Catt, J.A. 1985. Thermoluminescence dating of soils developed in late Devensian loess at Pegwell Bay, Kent. *Journal of Soil Science* 36, 293-98.

Wood, P. 1957-8. Second excavation of the strip lynchets at Bishopstone, near Swindon, Wilts., June,

1955. *Wiltshire Archaeological and Natural History Magazine* LVII(206), 18-23.

Wood, P. and Whittington, G. 1958-9. The investigation in 1957 of strip lynchets north of the Vale of Pewsey. *Wiltshire Archaeological and Natural History Magazine* LVII(207), 163-71.

Wood, P. and Whittington, G. 1960. Further examination of strip lynchets north of the Vale of Pewsey in 1958. *Wiltshire Archaeological and Natural History Magazine* LVII(208), 322-38.

Yorston, R.M. 1990. Comment on estimating tillage effects on artifact distributions. *American Antiquity* 55, 594-98.

Yorston, R.M., Gaffney, V.L. and Reynolds, P.J. 1990. Simulation of artefact movement due to cultivation. *Journal of Archaeological Science* 17, 67-83.

Zvelebil, M. 1986. Mesolithic prelude and Neolithic revolution. In Zvelebil, M. (ed.) *Hunters in Transition: Mesolithic Societies of Temperate Eurasia and their Transition to Farming*, 5-15. Cambridge: Cambridge University Press.

Zvelebil, M. 1994. Plant use in the Mesolithic and its role in the transition to farming. *Proceedings of the Prehistoric Society* 60, 35-74.

Zvelebil, M. and Dolukhanov, P. 1991. The transition to farming in Eastern and Northern Europe. *Journal of World Prehistory* 5, 233-78.

www.ingramcontent.com/pod-product-compliance
Lightning Source LLC
LaVergne TN
LVHW071220210926
848482LV00015B/857
9781407309804